Agile Manufacturing

AGILE MANUFACTURING

Globally Customized Green Products

M. P. Chowdiah

Distinguished Professor & Hon. Director
Centre for Agile Manufacturing
Research and Development
Bangalore University

I.K. International Publishing House Pvt. Ltd.

NEW DELHI • BANGALORE

Published by
I.K. International Publishing House Pvt. Ltd.
S-25, Green Park Extension
Uphaar Cinema Market
New Delhi–110 016 (India)
E-mail : info@ikinternational.com
Website: www.ikbooks.com

ISBN 978-93-81141-58-8

10 9 8 7 6 5 4 3 2 1

Published by Krishan Makhijani for I.K. International Publishing House Pvt. Ltd., S-25, Green Park Extension, Uphaar Cinema Market, New Delhi–110 016 and Printed by Rekha Printers Pvt. Ltd., Okhla Industrial Area, Phase II, New Delhi–110 020.

Dedicated to My Parents
Late Smt. & Sri Kalyani Puttaiah

Foreword

Weakness in Indian manufacturing technology must be strengthened if the nation is to enjoy by raising living standards together with a strong, stable position in the international market. **"Making Things Better : Competing in Manufacturing"** dictates that Indian manufacturing technology must improve in everything from agility, product design, etc. to manufacturing, process development, refinement, innovation, etc. This certainly suggests some policy on the part of innovation of the government also.

The governments should consider an array of options such as supply chain management/axiomatic/robust designs, intelligent robot and computer vision, intelligent systems in design and manufacturing, rapid prototyping, e-manufacturing, etc. that are important tools for advanced manufacturing technology. I am happy that most of these are dealt with in this book. In fact, the state and central governments must increase funding for model manufacturing centres like Centre for Agile Manufacturing Research and Development, Centres of Advanced Manufacturing, Digital Manufacturing, etc. and provide more funds for state industrial extension services/incubators which in turn are responsible in transferring technologies developed by the centres and other research institutes to manufacturers making new technologies usable by SMEs also demonstrated advance production technologies and making short-term loan for robotics, transfer line and other forms of advanced manufacturing equipment to even small entrepreneurs.

In view of the varied customer requirements–which is evident from the present market scenario–one would expect the manufacturing science to be agile enough and capable of meeting even **one customer market**. In order to effectively apply such a concept, the situation calls for intelligent integration of these three elements of manufacturing, i.e., technology, management and workforce embedded with agility in all the three. This integration would not become a reality unless it is ably supported by the effective application of information management in terms of networked database.

I am happy that this book on Agile Manufacturing has focused the chapters in proper sequence to meet the desires and demands of present customers. I have all appreciation for the author who has earned profound knowledge to create this new book on this emerging new generation concept of agile manufacturing.

I wish the author every success in his life.

V. Krishnamurthy
Chairman
National Manufacturing Competitive Council,
New Delhi

Foreword

Agile Manufacturing is a revolutionary approach which opens up a new pathway in designing and implementing manufacturing systems using new and emerging technologies and organizational managements – that is, management, technology, workforce to achieve the goals of competitive advantage. It is a continuous driving enterprise involving a quantum leap in our current thinking about manufacturing systems; development of the needed capability to respond to the demands of highly customized products of high quality. Agile manufacturing calls for integration of management, highly skilled and knowledgeable methods to penetrate highly competitive and dynamic niche market environments to successfully achieve world-class performance, even outperforming the best of competitors. All these requirements are mutually interdependent and hence we need a formidable strategy not only to become agile but also to exploit and dominate that agility. We concurrently need new productive skills and knowledge-enhancing research-based technologies so crucial to support the futuristic manufacturing systems without any traditional barriers that would frustrate profitable deployment of the system for the greater use of people.

Dr. M.P. Chowdiah, Director, Centre for Agile Manufacturing R&D, the author of this book and a scholar of long standing in Mechanical Engineering, Foundry Technology and Agile Manufacturing Research and Development has been experimenting for years–backed up by his vast engineering experience in India and abroad and his contacts with experts from wide across the world in the realm of this new frontier of manufacturing levering their extensive knowledge through national and international conferences and live-workshops dwelling on the subject with profound dedication. He thus unequivocally justifies his professional right and privilege to pen and publish this excellent book.

Authentically utilizing the benefits of his continuous study and aggressively applying them in the practical sphere of competitive agile manufacturing, Dr. Chowdiah presents here this book to help the reader to explore this new competitive battleground, assimilate its imperative secrets and thus empower him to gain into its leading edge and witness elusive a new success that has kept all concerned challengingly elusive this long. This success, the author hopes, will ease the transition to a new and existing era of progress and prosperity in the field of manufacturing industries where the society as a whole has common stake. In this light, this contribution will perhaps serve as one of the most helpful books covering the entire spectrum of agile manufacturing which is destined to rule the business world.

Er. M.S. Zaheed
Chairman and Managing Director
Hindustan Machine Tools

AND

President,
IIPE (Indian Institute of
Production Engineers)

Preface

Advances in the field of science and technology inevitably culminate in manufacturing products that serve mankind. Manufacturing, therefore, is facing challenges like never before. As new materials emerge into the world to meet the stringent requirements for performing functions for which they are selected, newer and appropriate manufacturing processes must be thought of. Mankind has been witnessing these changes with a hope that it would benefit by such advances. Big strides in Information Technology have been growing has greatly contributing to the manufacturing scene. The core of agile manufacturing is integration of technology, management and workforce with information technology.

In these contexts today manufacturing does not just mean mere technology, but much more that. It is essential that the business of manufacturing should encompass many more components to make it flexible so that customer-oriented products could be manufactured without loss of time in a cost-effective manner and business houses can compete in the world market with products of international standards. Capability is not just the criterion but how quickly this capability can be earned fruitfully. Hence, manufacturing community eagerly looks forward to share knowledge for mutual benefit.

Agile manufacturing is still in the infancy in many parts of the world. It is very appropriate that the Bangalore University hosted the International Conference on Agile Manufacturing, (ICAM 96), Liberalization has brought into the Indian scene a sense of discipline and competitiveness. It is heartening that an academic institution took up the responsibility of exposing the industry to the concepts and advantages of agile manufacturing. Industry houses, which are busy, as ever, in their own world should adopt the concepts of agile manufacturing not only to face immediate challenges but also the future.

M.P. Chowdiah

TEMPLE OF LEARNING

M.P. Chowdiah

- Studied B E (Mech Engg) at U V CE.,, (1951-1955)
- Went under Indo-German Scholarship for Training under 3 Companies (1956-1959)
- Worked as lecturer at NIT, SURATHKAL, (1959-1961)
- Completed MSME & Ph.D, PURDUE University, USA (1961-1965) and a MIT USA as P.G. Fellow (1965-1966)
- Worked as Project Engineer in GENERAL MOTORS Co; Saginaw, USA (1966-1971)
- Worked as Professor, DEAN & HOD at U V C E, Bangalore University (1971-1991)
- Worked as Hon. Director, & Distinguished Prof. Centre for Agile Manufacturing R & D, B U (1996-to Today)

Secy. & Chairman, The Institute of Engineers (India), Karnataka State Centre (1978-1988)
Chairman, National Design Research Forum (1991-1994)
Chairman, All India Council for Technical Education, SWRC (1994-1997)
President, (Engineering Sciences), Indian Science Congress Association, Kolkata
UGC Emeritus Professor (1997-2000)
President, Institute of Engineers (India), Kolkata (1999-2000)
Hon. Director, P.G. School of Studies, NITF, Peenya, Bangalore (1992-to Today)

Acknowledgements

I am immensely pleased to acknowledge the yeomen professional and technical services rendered by Shri B.R. Subramanyam who was Director and Editor (Technical) and Secretary and Director-General (I/c) of the Institution of Engineers (India), Calcutta for years with distinction. His outstanding performance in the correction and editing of the manuscript of this book has rendered the book a unique look. The invaluable data collected from various inter networks and libraries along with my knowledge and experience earned for 50 years has made its contents rich, useful and a storehouse of most up-to-date information in the subject area of Agile Manufacturing. I acknowledge the timely help and advice he gave me from time to time.

I acknowledge with thanks to Dr. G. Gargesa and Dr. V. Arun Kumar for their help in making a grand success of three international conferences:

1. "Agile Manufacturing" 22–24, February 1996, at Bangalore.
2. "Information Technology Integration for Manufacturing" (ITIM-98), 28-30, December 1998, at Bangalore, and
3. "Agile Manufacturing for Cleaner Products and Sustainability – Internet Solutions" 15-17, December 2002, at Bangalore.

I acknowledge with thanks the work done by Shri Purushotham who typed out almost all the chapters of the book, notwithstanding the huge number of corrections as and when part manuscripts were finished. I cannot forget Ku. Rajalakshmi V.S. who typed final copy with great care and concentration, before it was finally accepted by the publishers.

I acknowledge the untiring help given by my beloved wife Mrs. Sundra Chowdiah, who persuaded and encouraged me at every step to complete this book, without which I could not have completed the book in time.

My thanks to all those who have helped me with timely suggestions, specially Shri B. Raghunatha Rao, Former Sr. Scientist, CMTI, Bangalore and Dr. S. Vinodh, Professor, NIT Tiruchirapalli who have taken the pains in correcting the draft of this book.

My life becomes invalid, if I do not remember and express my deep gratitude to my mentors and nodal personalities of my life.

1. Late Dr. K.V. Puttappa who advised, compelled me and sent me to engineering. If I am something today, it is because of him. I am deeply indebted to him.
2. Late Swami Shambhavanandji, President, Sri Ramakrishna Ashram, Mysore was my spiritual guru. He was responsible for my going to Germany in 1958 under Indo-German Scholarship of India. I am always grateful to Swamiji.
3. Last, but not the least Sir M. Visvesvarya, an engineering wizard in the world, whose biography inspired me that I must dedicate myself to the service of the country. I am quite sure that his biography must have inspired hundreds and thousands.

I remember and worship all of them everyday.

M.P. CHOWDIAH

Contents

List of Figures and Tables

Abbreviations/Expansions

AIMA	All India Management Association	12.1.4
AMA	Agile Manufacturing Approach	3.2.1
AMEF	Agile Manufacturing Enterprise Forum	11.2.1
AMR	Advanced Managing Research	10.3
APS	Advanced Planning & Scheduling	17.3
AVIS	Automatic Visual Inspection System	15.7
BOM	Bill Of Materials	17.3
BPR	Business Process Re-Engineering	12.2
CAD	Computer Aided Design	12
CALS	Computer Acquisition & Logistic Support (Earlier Expansion) Continuous Acquisition & Life Cycle Support (Present Expansion)	4
CAM	Computer Aided Manufacturing	
CAPC	Computer Aided Production Planning & Control	12.2.2
CAPP	Computer Aided Process Planning	12.2.2
CAQC	Computer Aided Quality Control	20
CE	Concurrent Engineering	
CERP	Customized Enterprise Resource Planning	17.1
C & C	Computers & Communications	3.2.2
CIM	Computer Integrated Manufacturing	13.1
CNC	Computer Numerically Control	1.6
COViRDS	Conceptual Design Virtual Design Systems	15.8
CPPCS	Customized Production Planning, Control & Scheduling	17.1
CRP	Capacity Requirement Plan	17.3
CSRP	Customer Synchronized Resource Planning	17.1
CTS	Continuous Transition System	11.2.1
DFA	Design For Assembly	12.2.2
DFAM	Design For Automation	15.6
DFM	Design For Manufacturing	12.2.2
DFMA	Design For Manufacturing and Assembly	12.2.2
DFMEA	Design for Failure Mode Effect Analysis	12.2.9
DMU	Digital Mock-Ups	15.8
DSS	Decision Support System	20
DTC	Design To Cost	12.2.9
ECOs	Engineering Change Orders	1.1

EMOs	Exchange Messaging Outlook Software	20
EOQ	Economic Order Quantity	5.3
ERP	Enterprise Resource Planning	17.1
ES	Expert Systems	15.8
FEA	Finite Element Analysis	12.2.9
FMEA	Failure Mode Effect Analysis	15.6
G Flops	Giga Flops (Billion Calculations/Sec)	3.2.2
GDP	Gross Domestic Product	12.1
GMDH	Group Method of Data Handling	19
GNP	Gross National Product	9.11
GT	Group Technology	15.4
HRD	Human Resource Development	
HRM	Human Resource Management	21
IPPD	Integrated Product & Process Development	4, 18
IT	Information Technology	
JIT	Just-In-Time	12.1.9
KBES	Knowledge Based Expert System	20
LDF	Linear Discriminate Function	19
MBO	Managing Business Opportunity	12.1.7
MIPS	Million Instructions Per Second	3.2.2
MLT	Manufacturing Lead Time	15.9, 20
MMS	Man Machine System	20
MNC	Multi-National Companies	9.13
MRO	Maintenance Repair and Operations	20
MRP	Material Requirement Planning	17.3
MPS	Master Production Schedule	17.3
NPC	National Productivity Council	12.1.15
NTPC	National Thermal Power Corporation	12.1.4
OD	Organizational Development	12.1.7
OTC	One Touch Change Over	15.5
PDPC	Process Decision Program Chart	12.1.17
PIS	Productivity Information System	19
QAP	Quality Assurance Programme	19
QC	Quality Control	
QFD	Quality Function Deployment	1.2, 15.6
QWL	Quality Working Levels	12.1.7
RAL	Robotics & Automation Laboratory	15.6

RAPID	Rapid Product Innovation & Development	
RCCP	Rough Cut Capacity Plan	17.3
RPT	Rapid Proto Typing	12, 20
SL	Stereo Lithography	15.8
SME	Small & Medium Enterprises	12.1.4
SMED	Single Minute Exchange of Diagram	15.5
SMI	Small and Medium Size Industries	15.8, 20
SMLEs	Small, Medium and Large Enterprises	
SPC	Statistical Process Control	15.3
SPM	Strategic Productivity Management	
SQC	Statistical Quality Control	12.1.15
STL (file)	Stereo Lithography (File)	15.8 20
SWOT	Strength, Weakness, Opportunity, Threat	
TFD	Total Factor Productivity	
TPM	Total Productivity Management/Maintenance	15.8, 20
TQFD	Total Quality Function Deployment	15.6
TQM	Total Quality Management	9.13, 15.6
VAR	Value Added Reseller	15.8
VCR	Video Cassette Recorder	9.12
VFD	Variable Frequency Drive	20
VR-CAD	Virtual Reality Based Computer Aided Design	15.8
WIP	Work In Process	15.5, 15.9, 20

1

Products Development: The New Imperative

In earlier years, mass producers were very much dependent on invention organizations to create new products, and invention organizations were often dependent on mass producers to provide a ready market for their highly differentiated creations. This synergy between mass production and invention organizations worked very well for a long time. It was at the heart of America's economic success during the 20th century. But Japanese companies (in particular but not exclusively) figured out that if they continually improved their processes, they could achieve both lower costs and higher quality than the typical mass producer. By embracing a dynamic process change, they could gain a significant advantage over their competitors. This was so different from the old ways of doing things that it took American producers a long time to figure out exactly what it was. Now most companies subscribe to the continuous improvement business model (at least in principle, if not fully in practice), incorporating such innovations as statistical process control, cross-functional teams, total quality management, and customer satisfaction measurements, to name just a process life cycle of one execution. Every execution is different and better than the last.

While companies everywhere seem to be making great strides in quality by focusing on continuous improvement, many firms are already moving beyond this to the mass customization business model. Stable but very flexible and responsive processes provide dynamic flow of products, enabling companies to achieve both low costs and high individual customization. In this business model, the organization's primary thrust is to identify and fulfil individual wants and needs of every customer within the framework of this product varieties as configurations. Ideally, the product life cycle is that of one unit only as every product is different from the last and uniquely suited to particular customer's needs.

The first step in any product development is to methodologically determine which product to develop. Companies must realize that this is an important step and deserves significant effort proportional to its ultimate effect. The good news is that mass customization need not be a repetition of the short-lived competitive-advantage games.

1.1 BUILD TO ORDER

1.1.1 Elimination of Carry Over Inventory Cost

Mass customized products are built to order in flexible plants. This capability to build to order is a key factor, minimizing the cost that will allow mass customized products to compete with mass produced products and yet provide the additional competitive advantage of exactly meeting the needs of individual customers.

The advantage is that factory finished goods/inventory is eliminated by building products to order, instead of building to forecast and then holding products in a warehouse until ordered by distributors or customers. Like work in process (WIP) inventory, finished goods inventory may cost the same to carry except that finished goods are completed and therefore are more valuable. Using 25 per cent of value per year, $10 million worth of finished goods in inventory would cost $2.5 million per year to carry. Build to order methods can eliminate factory finished goods inventory and, thus, save its yearly inventory carrying cost.

Dealer finished goods inventory can be eliminated if customized orders can be quickly fulfilled and delivered to the customer. As with factory inventory, dealer inventory also has a carrying cost. Even if the dealer/distributor is separate from the manufacturer, the carrying cost will have to be paid ultimately by the customer. For example, an automobile dealer with 200 vehicles (Nissan Make) in stock for one year with an average value of $20000 each would represent $4 million worth of inventory. Using an yearly carrying cost of 25 per cent, the carrying cost would be $1 million per year.

When new car prices exceed customers ability to afford them, the customers buy more used cars as a cost-effective alternative. Built-to-order new automobiles could compete well against used cars, with lower prices, since the new cars could avoid dealer inventory expenses, whereas used cars must be stocked in inventory.

Supply chain inventory can be minimized since build-to-order products do not need to be stocked at various warehouses along the supply chain: at distributors, consolidators, forwarders.

1.1.2 Rationalization and Standardization

Before starting the design of mass customized products, it is important to make sure that all the prerequisites are in place or will be in place by the time the products go into production. Specifically related products should be rationalized. Standardization programmes should be well under way as well as improvements in manufacturing flexibility to lower the cost of variety, all decisions should be made on the basis of total cost, and product development itself must become an important core competency.

1.1.3 Translating Customer's Voice to Product Design Specification

The first step in any product development is to translate the voice of the customer into product design specifications and resource prioritizations. For mass customization, this step is even more important and more complex since the voice of the customer must be translated into families of products that define the range of customizability, needed to satisfy customers.

Contrary to the common temptation, the first step is not concept work. But unfortunately, product development is often launched because of intoxicating concepts that may or may not satisfy customers.

1.1.4 Product Definition

In a survey of 12 large technology-based companies, managers were asked, 'What are the causes of product development delays'? The number one response, with 71 per cent of managers responding, was poor product definition. If the product is not defined well, then the product definition may have to change later to really satisfy customers. Late changes in product definition usually result in costly and time-consuming engineering change orders (ECOs) because of the amount of engineering and possibly tooling that have been based on the original product definition. An equally unpleasant alternative to late changes in the product definitions is to proceed based on the poor product definition and introduce the product that does not really satisfy customers.

A good product definition will become a consistent focus for the whole company synergy that is the source of competitive advantage in product development. One company waited until the prototype stage. Since that was not what it wanted, the result was much rework to redesign the product in order to reflect what was wanted.

The proper procedure for an intriguing concept or technological strength would be first to thoroughly investigate the market potential for mass customization—the scope and range of customization—the universe of possibilities which need to be optimized for the greatest potential of success in the marketplace without wasting effort for unnecessary customization.

1.2 AN OVERVIEW OF QUALITY FUNCTION DEPLOYMENT (QFD)

QFD is a tool for systematically translating the voice of the customer into product design specifications and resource prioritizations. Its strength is to translate subjective customer wants and needs into objective specifications that engineers can use to design products.

In the most general sense, the input to QFD is a set of subjective customer preferences; the outputs are product specifications and resource prioritization. The values in the 'design specifications' are the actual values which the engineers will use to design products. Resource prioritization is the per cent of the design team's effort that should be spent on each aspect of the design. Using this, prioritization will ensure that the design team devotes its efforts toward features that customers want the most. And, thus, they will not waste effort in 'polishing the ruby' more than the customer wants it polished. This can be a temptation when engineers are personally excited about certain new technologies that are being used in the product.

1.3 PRODUCT PLANNING AND CUSTOMERS NEED FULFILMENT

Early actions in product development are crucial to ultimate success and exert significant influence on cost and time to market. Unfortunately, despite the importance of these early steps, too many product developers almost bypass these early steps by making snap decisions based on whims, limited customer exposures, intra-company politics, competitive precedents, or the excitement decision, a decision to pursue the status quo or blindly accept some 'default' choices.

1.4 PROCESS INFRASTRUCTURE

Modularity, as mentioned earlier, when discussing the movement from continuous improvement to mass customization, applies not only to products but also to the processes by which they are manufactured. Product developers must work with manufacturing engineers and operators to create a process infrastructure that automates the linkages between modular process capabilities—atomic units of value that the company knows how to efficiently perform. Each of the exemplars discussed earlier employs such an infrastructure that enables it to be as effective as possible in ascertaining what customers need, in linking customers need—unique capabilities as effortlessly as possible.

Many R&D functions implement groupware to share information and resources. While that can be a very good start at breaking down the walls of communication that have grown up around the 'Not Invented Here' syndrome, today's groupware offerings in no way fill the needs required of process infrastructures. Groupware, at least as it has thus far been implemented, is suitable only for invention, not mass customization, because it is fundamentally based on ad hoc activities. Existing tools have no notion of process capability modules that are essentially resting in inventory, waiting to be called upon to fulfil the next customer's wants and needs. Rather, what a company needs are work flow management, coordination software, enterprise networks, common customer views (such as with image processing or shared databases) computer-integrated manufacturing and agile manufacturing capabilities—what are generically categorized here as process infrastructures—that can all be used to automate the linkages between processes and the relationships between people, enabling exactly the right resources to be brought together to cost-effectively service a customer's unique wants and needs.

Creating such an infrastructure requires a significant financial investment and implementation effort. This may be difficult—if not impossible—to cost-justify by current financial procedures because it is impossible to forecast exactly what process capabilities customers will want at any particular time. While academia and consultants are making progress on how to do this better (such as strategic options theory), when it comes right down to it, companies thus far have had to make a commitment to mass customization based on faith (that's right: faith!) in their own ability to create the future, rather than their own perception of what worked in the past. Top management must be convinced that the future will be a battle over who can most efficiently serve each customer uniquely or there may not be enough of a future to worry about.

Development of learning relationships and their benefits

- **Increased margins:** Because products are tailored precisely to customer needs, customers receive greater value and are thus willing to pay a premium price.
- **Increased revenue per customer:** Because the company knows more about customers than any competitor, they keep coming back every time they are in the market for what you offer.
- **Increased number of customers (at lower acquisition costs):** Because customers are pleased with the experience, they will tell their friends and associates, many of whom would try it and join in; in turn, they would tell others, and so on.
- **Increased customers retention:** The more each customer learns about his wants, needs and preferences, the more difficult it will be for him to obtain an equivalent level of service from a competitor.

1.5 RENEWAL OF RELATIONSHIP

Learning relationship enables companies to essentially keep their customers forever with two provisions. Firstly, once in such a 'monopoly' relationship the company must not jack up its prices.

Secondly, the company must continue to learn from its customers and fulfil what it learns, more precisely tailored products. This, in turn, requires that the company must be able to renew its product architecture and process infrastructure in response to specific customer requirements that cannot yet be met. This means 'leaving' the mass customization model and returning to invention to create something truly new.

This does not mean that every customer's whims must be satisfied, or that every real requirement must be fulfilled. However, a company should determine what to do with every requirement through a decision-making process as below:

(i) Does this fit within the universe of possibility that is defined? If not, it should be fulfilled. Track it along with all other unfulfilled requests to determine when the present product architecture should be abandoned in favour of a more encompassing one.
(ii) If so, does this make sense to do right now? Consider such factors as the lifetime value of the customer and how much this decision will hurt or help the relationship and the willingness of the customer to help pay for the development; which other customers would desire this functionality, how much time and costs are involved, and what is the current capacity of the organization for renewal. If not, schedule it for a later day or for re-examination at some point, again tracking the requirements for further consideration.
(iii) If so, invent the module for this particular customer. And then determine whether this module should 'remain' in invention or should be developed, linked, and modularized to be added as a standard module to the product architecture or a standard capability to the process infrastructure. Many of the same factors as cited in point (ii) above come into play here.

When faced with new customer demands, many product development groups will seek to invent new products only when many requirements are known and understood. The spirit of mass customization is that the modular architecture and/or infrastructure enables this to be done incrementally, one module at a time, at very low costs. Instead of massive development efforts—which often end up with a large share of the time spent reinventing what someone else somewhere has already done—new modules can be created and dynamically linked into an existing architecture with a fraction of the effort. This is exactly what Motorola does so well with its Fusion Factory, or what Ross does everytime it works with a customer to determine how its pneumatic valves can enable that customer to realize strategic value. They do not completely reinvent a new pager or a new valve; they work from existing designs to reuse 80 to 90 per cent (or even 95 or 99 per cent) of what they have used before, and then spend their truly creative work only on what has to be innovated to meet that particular customer's needs.

1.6 MASS CUSTOMIZATION VISION AND ITS MODEL

Few products today can be manufactured like mass-produced items without any need for product variation or customization. Most manufacturers are forced to customize products to some degree for increasingly selective customers or to compete in niche markets. However, if products are not designed well for this and if manufacturing is not flexible enough, then customizing products will be a slow and costly ordeal.

The next paradigm, after the century-old mass customization, is mass customization with the ability to design and manufacture customized products at mass production efficiency and speed. This will enable to develop products that can be quickly and inexpensively customized for niche markets as well as for individual customers.

Success will depend on strong product development methodologies that enable multifunctional design teams to concurrently design whole families of products and flexible manufacturing processes. Products must be designed around common parts, versatile, standardized interfaces, common fixturing geometrics, and standard processes.

To assure manufacturing flexibility, products and processes must concurrently be designed to eliminate all setup steps, such as kitting, retrieving parts or tools, changing fixtures, manually positioning parts or finding instructions.

The results will be well worth the investment. Companies that use these techniques will experience ultra-fast development of modular products, the ability to quickly respond to changing markets and opportunities, delivery state, superiority, price premium opportunities, lower overhead costs and better satisfaction of customer needs.

Mass customization can be implemented by way of an example of a product like an electronic system. This generic category includes all types of computers, instruments, communication devices, audio-visual equipment, electronic games, small appliances, and so on. Specially, this example is based on a small instrument that can be customized with unique software and many different combinations of meters, dials, cases, and various internal parts. This model is structured to be very doable in that it does not depend on high levels of automation or high volume. The only programmable machine tools used are circuit board assembly equipment and CNC machining centres and most common CNC machine tools in industry. The final assembly is manual, which is selected for this example for several reasons. Most companies may not have the volume or the budget for the equipment necessary for automated mass customization. Of course, all the principles demonstrated here can be applied to more automated operations. The whole operation is build-to-order with Kanban part re-supply which can be independent of MRP systems, which are very much dependent on reliable forecasts and accurate bills of materials.

Almost all product lines could use some rationalization, which is systematic analysis and consolidation of product lines to align them to achieve long-term corporate goals. Its purpose is to identify products that do not fit into a flexible environment, have low sales, have excessive overhead demands, are not really appreciated by customers; or have limited future potential or may be losing money. Rationalization analysis may direct companies to drop these unprofitable products; or if 'completeness' is important, farm out their manufacture to others when they are not within company core competencies.

This rationalization process should also be applied to product options and variations. Many companies structure bills of materials so that every unique combination of options is a 'product' regardless of how they are defined, options may have escalated and consequently, could benefit from some scrutiny. The case of Nissan has already been cited earlier to reduce proliferation.

One solution to reduce proliferation is to eliminate the low selling options and to make the popular options of standard equipment or combine appropriate sets of options into packages. The companies must avoid the temptation to base rationalizing decisions on reported profitability.

Companies may think that they automatically rationalize products by routinely eliminating products that 'drop below the line' of adequate profitability, as reported by their product costing systems. However, if product costing generates enormous distortions, then this automatic rationalization may not have occurred, products that were really profitable may have been erroneously eliminated which is the worst thing to happen.

1.7 AGILE PRODUCT DEVELOPMENT FOR MASS CUSTOMIZATION

The phrase 'agile product development' can be interpreted in two ways, both of which are correct and applicable to a wide range of products and industries:

(i) An agile product development process that can rapidly introduce a steady succession of incremental product improvements—which can be called 'new' products—that are really planned 'variations on a theme', based on common parts and modular product architecture. This capability results in ultra fast time to market, much faster than possible with independent products that do not benefit from product-family synergies in design and manufacture.

(ii) Development of agile products that can be manufactured in any of the following agile environments: Agile/Flexible/Continuous–Flow Manufacturing, Just-in-time, Build-to-order, and Mass Customization.

Mass customization requires a strong product development methodology infrastructure.

2
Standardization: Steps and Prerequisites

2.1 RATIONALIZATION CRITERIA

It is important to rationalize the production line before implementing mass customization or any of the agile paradigms: just-in-time, build-to-order, flexible manufacturing, continuous flow manufacturing, or niche market competition. This step would be parallel to simplifying product designs before automating assembly, and simplifying the information structures before computerizing. Failure to perform these basic prerequisites leads to the industrial equivalent of 'paving the cow paths'.

Rationalizing product lines may improve company profits by eliminating and/or outsourcing products that are currently depressing company profits. It will also lead to better understanding of company core competencies, help prioritize continuous improvement activities and facilitate better strategic planning.

When rationalizing products, one has to be sure to rationalize families of products rather than single products. Families of products do not necessarily share the same page in the catalogue, but do share some similar products; hence they could be considered together if their processing is almost the same with some minor differences that have little cost or throughput consequences.

The following techniques can be used to rationalize product lines for existing products. The techniques are presented in order of the most objective first and the most subjective last.

1. Sales volume–plot all products in Pareto order
2. Sales revenue–plot all products in Pareto order
3. Part commonality–plot products by percentage of common parts
4. Cost of variety–plot all products
5. True profitability–plot all products
6. Polls and surveys–plot responses
7. Factory processing
8. Functionality
9. Customer needs
10. Core competencies
11. Clean-sheet-of-paper scenario
12. Future potential

3

Globalization of Product Design

3.1 STEPS TOWARDS JUST-IN-TIME PRODUCTION

The globalization impact on product design is an integration of the product design policy with equivalent stages of manufacturing, education and practice for customized products with quality assurance. Product quality assurance design demands quality at all stages of material selection, the selected design, established processes and fabrication, assembly, testing, packing and finally marketing.

To execute the standard product design, the workforce must be injected with quality education based on source of control such as admission criteria, curriculum design, programme selection and implementation, evaluation and employability, etc.

Since economic reforms based on policy of globalization affect the industrial and economic scene by increasing competition in the quality and production of the products and services, they must prove to be of highest quality and yet competitive. It is therefore necessary to have a desired policy of research, development and technology transference.

3.2 HISTORY OF MANUFACTURING

Before discussing the changes in manufacturing for globalization, it is necessary to discuss the history of manufacturing from 1942-2000. The sequential scenario of manufacturing in the world presents the history of manufacturing from the past to the present status.

The impact of globalization on manufacturing has led to **AM**, 'Agile Manufacturing which may be defined as the science of a business system that integrates management, technology and workforce, making the system flexible enough for a manufacturer to switch over from one component that is being produced to another component that is desired to be produced in a cost-effective manner, in a short-time, within the framework of the system'.

3.2.1 Look Out for Further Progress

Looking back on human affairs from a technological point of view, we see that man has experienced three major transformations.

- The first was the **agricultural revolution**, characterized by the transition from a society based on hunting and gathering, to one based on cultivation, which made it possible to

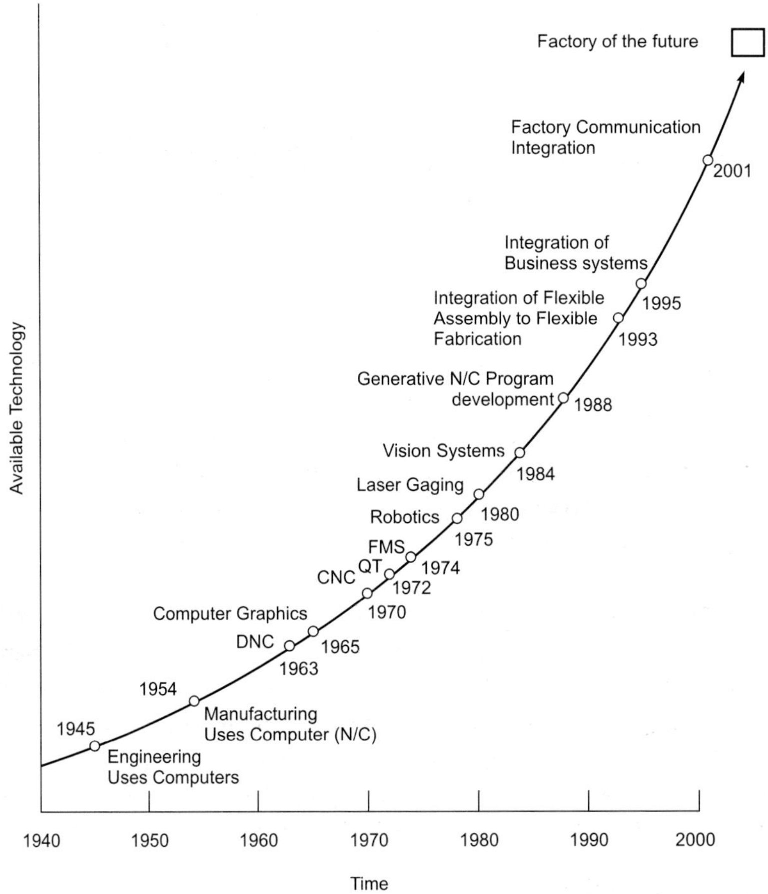

FIGURE 3.1: History of manufacturing from 1942 to 2000.

sustain large populations. In Japan, this change occurred during the Roman Period (around 2500 BC) by which time the population had reached about ten times its former level.

- The second great transformation was the Industrial Revolution. This event began in England in the 18th century, spreading to other parts of the globe and reaching Japan in the late 19th to early 20th centuries. Its inception was the invention of the steam engine, followed by the internal combustion engine, electric power, steel, chemicals and many other innovative technologies. Parallel to this, the population of the world, and of Japan, grew exponentially.
- The third great transformation is what has been faced today by mankind, brought out by **Information Technology**, Biotechnology, new materials technology, energy technologies such as nuclear power and space exploration. As in the industrial revolution, there is no single technology which by itself constructs a new social order. There is, however, a core technology here: that of information.

Information technology has provided us with an increased intellectual capacity that was inconceivable in the industrial revolution. It is being integrated with other technical areas: with biotechnology to produce bioelectronics: with new materials technology to become, for example, optoelectronics; with nuclear energy technology to become laser generated nuclear fusion. These technologies are greatly benefiting mankind, bringing a shift in emphasis from quantity to quality. The previous two transformations engendered great leaps in population but the current change will foster qualitative fulfilment for mankind.

Information Technology is the backbone–the cornerstone of a new social order and is the backbone in integrating C&C. Indeed, computer and communication technologies have already begun to merge, aided by the rapid progress that has been made in digital, semi-conductor and integrated-circuit technology. Thanks to digitalization, present levels of computing power are 100-200 MIPS (million instructions per second) for general-purpose computers and two GFLOPS (gigaflops, i.e., two billion calculations per second) for supercomputers; communication technology will soon reach a point where 1.6 billion signals per second can be transmitted.

Semiconductor technology, developed at Bell Laboratories in 1948, has also seen remarkable progress. The amount of memory capacity, we can store on a five-millimetre square silicon chip quadruples every two years. Communication circuitry too can be placed on the same chip. This enables the creation of small computers with communication capability, facilitating computerized networks. These developments have made possible distributed information processing systems and one of the common manifestations of C&C. Yet another such manifestation of C&C is the fifth-generation computer, which takes advantage of enormous chip capabilities.

Progress in hardware technology will continue at a rapid pace that will, in fact, accelerate due to such developments as bioelectronics and new materials. Further research has resulted in numerous breakthroughs, such as new angstrom-level devices that permit molecular control and the biosensor which will pave the way for bioelectronics. Particularly noteworthy is a material that attains fiftyfold performance improvement through trial manufacture in space. And that is glass for optical fibres which was made in the space shuttle. When this glass was tested, it was found that it was transmitting light 10,000 kilometres without the use of repeaters. If space factories which use the advantages of low gravity and vacuum in their manufacturing processes permit the creation of new and high performance materials which is not possible to make on the earth, hardware technology will take yet another great leap forward. Those who dominate materials will dominate technology. It is vital, therefore, not to underestimate the importance of hardware.

Today, those who dominate software dominate the world, for software is of inestimable importance in bridging the communication gap between man and machine. At present, extensive training is often required before machines can be operated. In the future, however, Artificial Intelligence will eventually overcome this problem. For example, artificial "ears" will be able to understand correctly even heavily accented voices, enabling machines to move and operate according to the speaker's wishes.

3.2.2 The New Competitive Framework

As a result of all these propitious combinations, the new competitive standards we are now facing are like birds of a feather. That is, they are mutually reinforcing and developing simultaneously. They are understood best as a framework in which each standard makes sense only in the context of the others. Individual employers who begin by emphasizing one of those standards usually end up emblazing them all because each standard overlaps and leads on to the next. In the final analysis, the distinction among the standards is in part semantic. Each standard is integrally connected with the others in a flexible and organic whole.

4

Manufacturing Competitiveness: The New Phase

Manufacturing competitiveness and "world-class manufacturing" are often used interchangeably. Establishing manufacturing competitiveness is considered equivalent to being a world leader in "world-class manufacturing". In other words, manufacturing competitiveness means sustained growth and earning through building customer loyalty by creating high value products in very dynamic global markets. A number of modernization programmes have been launched by many companies, such as FORD, GM, IBM, NEC and Newport News Shipbuilding dealing with how a company does business (Prasad, 1994). These programmes focus on business processes and information usage in the manufacturing environment. Programmes such as enterprise reengineering, multidimensional integration of engineering, and manufacturing and logistics practices have resulted in far reaching changes and profits. With changes, we are witnessing a movement from the "age of control" to the "age of flexibility". These changes describe a set of dominant trends in the business environment that have influenced the competitiveness of companies since 1980 (Goldhar et al. 1991). These changes could not have been possible without process tools like concurrent engineering (CE), and Continuous Acquisition and Life Cycle Support (CALS–earlier known as "Computer-Aided Acquisition and Logistics support") in action (Bauman, 1990). Attributes such as empowerment, flexibility, quality, agility, fast-to market, accountability, teamwork, and interaction are inherent in the notion of CE.

Concurrent engineering is considered one of the key concepts that enables companies to attain world-class stature (Shina, 1991). Today, it is an organizational keyword. The term refers to the same organizational process described as simultaneous engineering, life cycle engineering, parallel engineering, and sometimes as the multidisciplinary team approach or Integrated Product and Process Development (IPPD) (Hall, 1991; Siemake and Spann, 1991). Here, everyone contributing to the final product, from conceptual design to marketing earns, is required to participate in the project from its very inception. Some industrial analysts claim that this systematic approach is revolutionizing the manufacturing industries and could be just the boost that will enable firms to survive in the competitive global marketplace. Others argue that this practice is not really new and that it is very similar to the old fashioned "team work" that was prevalent a decade ago during product design and manufacturing (Dida and Begley, 1991). A decade ago, the size of the project teams was small compared to today's standards (a couple of dozen people at most). It worked reasonably well, because small teams can work more closely than large teams. There are numerous anecdotes describing such small teams that expose challenging issues and resolve them by mutual understanding (Chisholm, 1973). However, when project sizes became larger (involving hundreds of people), large teams paradigms have been in a dynamic state of flux for some time. In order to understand the changing pattern, let us review the situation by taking a closer look at the historical manufacturing processes and practices.

- Products were designed and built to customer specifications.
- Each craftworker paid attention to details, quality was built-in-by design and in-process quality control was the responsibility of each worker.
- No in-process inventory was required, materials were purchased just-in-time, and finished products were delivered on schedule and on cost.
- Designers knew well in advance the manufacturing requirements including the capabilities and the limitations of the production process; they had the power to modify in-process design (there was no need for empowerment) to take advantage of changing situations.
- Craftworkers participated in all decision-making. They worked concurrently, switched roles, and helped each other.

In craft manufacturing, since the product was crafted manually the following cases occurred:

- A batch size, or sometimes a lot size of even 1 was used.
- Productivity was limited to the capacity of a few craftworkers and their skill ability.
- Most costs were tied to production labours.
- Operations were sequential in nature.
- There were no major incentives to control costs as profitability was cost centred.

4.1 DEVELOPMENTS DURING PAST ERAS: MASS PRODUCTION 1965 + (1)

During this period (1960-70), consumers demanded more and more functionality from products, especially durable goods. The concept of "mass production" took off. This evolution of competitive advantage via "economy of scale" was the second reach. Mechanization (i.e., use of machine power) was sought to efficiently build, fabricate, and assemble components (automation). The manufacturing process still consisted of discrete serial steps along the line of crafts manufacturing. However, the economies of skills was no longer the equation for gaining competitive advantage in the manufacturing world.

The skill sets for production of goods changed from a handful of people to a multidisciplinary team of experts. Developing a product required assistance of a number of experts and other individuals experienced in the many aspects of the product's life cycle, such as design, manufacturing, sales, product support, marketing, etc. Internal industrial emphasis changed from engineering and production to include marketing and finance. Interest in the marketplace and investment grew. Buying existing industries became more lucrative than improving efficiency of the internal operations or practices.

4.2 AN ERA OF MECHANIZATION 1970 + (1)

During these ten years (1960-70), as industry moved toward more complex products and the automated assembly line, new ways to get uniform results were sought. Standards were created for precise drawings, interchangeable parts, tolerances, and quality control. Skilled employees were hired to cope with these demands. During the later part of 1970 a new emphasis was placed on mechanization—the ability to mass produce identical, interchangeable components. The "economy of scale" took the form of an "economy of machine power". This became the **third reach**. This reach employed a process strategy based on "mechanization" for reaching an adequate "economy of scale" for equivalent mechanical products. Mechanization not only became popular replacing human or animal power, but also infiltrated the assembly lines. Robots were sought for the production of complex assemblies and repetitive chores. Mechanization drastically changed the manufacturing economy of scale. Manufacturers of products made from dissimilar parts soon found it difficult to compete. Due to mechanization, worker's skills became highly specialized and job scopes became highly focused.

The marketplace is also rapidly changing. During the last decade or so, 80 per cent of the Fortune 1000 companies have moved to client/server mode of computing. The following describes some trends:

- The shortage of trained and talented workforce (including programmers) has increased.
- Information technology (IT) cost of hardware has declined.
- IT mix of computers and varieties (PCs, workstations, minis, micros, and brand names) has increased.
- The number of applications that are being run at the workplace has increased.
- The cost of the IT services has stayed fairly level.
- The global competition in the manufacturing sector has increased significantly.
- The computer speed and its power have increased.
- The integration cost of software has steadily climbed.
- Time-to-market has been steadily declining.
- Product complexity has increased over the years.
- The software development cost has been rising exponentially.

The factors that demand a significant change in overall approach to design and manufacturing include the following.

- Rapid proliferation in the number and variety of products, we manufacture today.
- Increasing dependency of product performance on manufacturing process capabilities.
- Customer demands for improved quality and better product performance.
- Dependence of cost on labour-rates resulting from choice of plants to manufacture the product and their locations.
- Demand for stronger and lighter materials.
- Competitors steadily shortening their time to market.
- Availability of world-class quality products at competitive prices in the world market.
- Declining cost of computing hardware versus the power it offers.
- Rising costs of software development versus the hardware costs.

Today, customers expect the same level of satisfaction and perfection that initially came from craft manufacturing but they want to pay a price achievable only by mass production. Customers want the best of both worlds "features". Attention to details, and "quality" from craft manufacturing, and "low cost" and "consistency" from mass production are early sought for. Though, the underlying process and teamwork used in craft manufacturing represents a miniature replica of what is desirable for a customer-focused organization, it lacks the "recipe" for a successful company. The main question is how one would be able to provide customer satisfaction on a recurrent basis if technology and organization keep on changing. Chain competition has driven engineers to include terms such as time compression, total quality management, teamwork, quality function deployment, and Taguchi into their vocabulary (Carey, 1992). Concurrent engineering, coupled with automation efforts, is vital in maintaining and communication technology, it is still not an easy task to win competitiveness effectively. Despite all of the above efforts applied in the United States, foreign organizations appear to be gaining ground from the consumers' viewpoint. The "change" in external conditions adds new competitive pressure to further decrease the time and lower the costs. Today, companies are facing tremendous challenges of how to provide the agility that came from craft manufacturing with the cost benefits that were the results of mass production.

4.2.1 Agility (Economy of Ease or Flexibility)

A prestigious study at the Iacocca Institute of Lehigh University has defined agility as the paradigm for manufacturing in the next century (Goldman and Prices, 1991). Agility is defined as the ability to thrive and prosper in a competitive environment of continuous and unanticipated change, and to respond quickly to rapidly changing markets and customer demands. A key element of agility is the agile enterprise. Agile enterprises are totally integrated organizations where information seamlessly flows among manufacturing, engineering, marketing, purchasing, finance, inventory, sales, and research (Dove 1993). Just as mass production 1950 leveraged "machine intelligence" (economy of skills), agile manufacturing is the "economy of ease" with which a company can react to new opportunities. Becoming an agile, virtual company does not necessarily mean downsizing. "Economy of ease" is taking the front seat as a fifth reach. It is not a technological solution. If plugged in, it will enable an enterprise to respond to change. It is merely a concept to be used while designing products and manufacturing processes so that when markets change or when customers need change, agility in the process can come to the rescue, allowing reconfiguration of the setup with calculated penalty and change in product offerings. Agility is part of the measure of merit that needs to be considered in product realization.

4.2.2 Throughput

A global organization requires global or integrated thinking. The former CAD measures of productivity, such as "engineering CAD throughput" or the number of drawings created in a week are (thus inadequate) throughput measures. It is not a true representation for maximizing the flow of work. Measures on "number of engineering changes per unit of time", or "ramp-up-time-to-volume production" are closer to global measures. Such measures on work-flow provide a more accurate big picture. Throughput is the rate (per unit of time) at which the current factory process transforms the materials into finished "useable" products. Useable means the measure does not include defective or scrap parts (that is, those that can be converted into sales or profit dollars). Clearly, throughput is dependent on the amount of lead time required to finish each one of the activities. Throughput for serial engineering is lowest, since one starts when the other ends. Throughput for concurrent engineering is proportional to the degree of overlap or independence between the activities. If each of the activities is completely independent, then throughput will be more because when activities are independent all activities can be started at once.

4.3 THE CHANGE IMPLICATIONS

The established practices that we now use in manufacturing industry were developed when circumstances were very different from those existing today. In the past, the manufacturing environments was largely cost driven and was based on high volume, low variety markets, with low levels of competition. In this environment, we found that management accounting techniques based on financial measures of performance, and management and manufacturing practices based on the application of division of labour, hierarchical control, centralization, de-skilling and so on worked reasonably well.

For example, when technology was applied to automate production and to de-skill people on the shop floor, generally it was found that this produced competitive advantage and made companies more profitable. Furthermore, failure to adequately address people's issues during the design of production plant was not seen as a source of competitive disadvantage. Neither did product development processes based on a serial engineering model (the over-the-wall approach) constituted a major competitive disadvantage, because most companies used this method and there was little or no incentive to improve the product development process.

The profound changes that are now occurring in the marketplace have already initiated major changes in the way we do things in manufacturing industry. Our belief that the introduction of new technology to reduce costs will guarantee improved competitiveness and profitability has been challenged by concepts such as just-in-time manufacturing, concurrent engineering, and so on. Furthermore, increased emphasis on quality and flexibility has meant that we have begun to realize their importance.

The following are the four core concepts underlying agile manufacturing—all four are mutually interdependent. We need to develop a strategy to become agile, but we also need a strategy to exploit our agility.

- To develop agile manufacturing, the strategy should be to focus on ways to achieve integration of organization, people and technology.
- To achieve integration of organization, people and technology we need to develop interdisciplinary design methodologies.
- It is the integration of organization, people and technology that delivers competitive advantage.
- There are many different forms of integration. Integration of technology is only one narrow view of integration.

4.4 COMPETITIVE ENVIRONMENT OF THE FUTURE

The nature of the changes that are occurring in the competitive environment is well described by members of the organization and efficiency department of Philips (International) in their important book, 'Flexible Manufacturing—Integrated Technological and Social Innovation' (Bolwijin, et al. 1986). These changes are concerned about shifting the emphasis away from the primary concern about price, to addressing simultaneously and without trade-off, issues of price, quality and customer choice.

4.5 DYNAMICS OF COMPETITIVENESS

Hayes, et al. (1988) have pointed out that we do not always fully appreciate our competitive position at any point of time is often less important than the rate of improvement of our competitive position as compared to our competitors. Manufacturing competitiveness is a dynamic process—it is not static. Rate of change in competitive performance needs, therefore, to be related to the rate of change of competitors. The competitive performance of world class performers is continually improving (a moving target). To catch up and overtake them, others must achieve a rapid change in competitive performance, and to stay ahead they must keep improving competitive performance at a faster rate than the competitors. If the rate of change is allowed to fall off to any degree, then it will only be a matter of time before the competitors take the lead again.

In order to understand how to maintain competitive advantage, it is important to appreciate the mechanism of innovation. From time to time, new discoveries lead to technological breakthroughs which have a revolutionary impact. These innovations spawn whole new industries, new products and new ways of doing things. Examples of revolutionary developments include the steam engine, the Bessemer blast furnace, the dynamo, telephone communications, manned flight, transistors and many others. However, all these technologies, once introduced, have been subject to on going improvements over a long period. None of the innovations, nevertheless, are a vitally important part of the innovation process.

Rapid changes in competitive performance are only likely to come about through revolutionary changes. This means that we have to get involved in making a big-bang; that is, major and highly innovative changes of this order (in terms of organization, people, technology, products and processes, and so on). Once a competitive edge has been achieved through such innovative changes, it is more likely to be sustained by smaller and more incremental changes involving continuous improvements to products, processes and organizations, people and technology. But we should also be looking to make revolutionary changes again at some point in the future, as our competitors will also be seeking to make rapid improvements in their competitive position.

Thus, in an innovation process, we experience revolutionary changes, but in between there is a process of incremental improvement. Although revolutionary breakthroughs happen from time to time, they are much rarer than popularly believed, and breakthroughs that are hailed as revolutionary often turn out to be less revolutionary than prophesied. The majority of improvements most often come from small, regular and continuous incremental improvements in all areas of the business through the cumulative effect of all these indeed.

The important point about the dynamics of competitiveness is that it is an interactive system. We must occasionally make revolutionary changes in order not to overtake us once again.

4.6 INTEGRATING MANAGEMENT, WORKFORCE AND TECHNOLOGY

Agile manufacturing is achieved through the integration of three primary resources into a coordinate, interdependent system. These three primary resources are: innovative management structures and organization, a skill base of knowledge and empowered people, and flexible and intelligent technologies. For brevity we refer to these simply as Organization, People and Technology.

Thus, an important feature of agile manufacturing is the philosophy that competitive advantage is not only derived from technology, but also from appropriate and responsive organizational structures, appropriate work practices, and from the skills, knowledge, intelligence, experience and creativity of all the people involved.

In agile manufacturing we are suggesting that appropriate organizational structures and people, with their unique abilities, actually provide us with improved competitiveness. In essence, therefore, the organization and people issues in manufacturing are not just a question of human factors, but are more a question of developing strategies that specifically set out to use organization, people and technology to achieve improved competitiveness (Kidd, 1991).

4.7 BALANCED MANUFACTURING

In the past, we have tended to approach the development of advanced manufacturing systems and the associated computer based technologies, as a technology-oriented activity. The direction that we should now take is based on a concept of balance. By this we mean that we should develop a new approach which integrates organization and people's issues into the manufacturing systems developments process.

Manufacturing systems design needs to incorporate modern organizational concepts and work practices. Instead of using technology only to improve business performance, we should also implement organizational innovation, and change our traditional work practices. The role of our production associated on the shop floor needs to be expanded and technologies are required to support all these changes. This type of thinking represents a sensible and balanced use of organization, people and technology. It is also good manufacturing systems engineering design practice.

Thus, our aim is not to shift from technology-oriented manufacturing to a purely human and organization-driven approach. We are proposing a balanced approach: one that gives serious consideration to the organization, people and technology. Our argument is that we need to take a broader view of manufacturing, always seeking the right mix and the right balance for each problem encountered. The three dimensions are complementary.

Balance is an important conceptual basis for the new paradigm of agile manufacturing. It will help us to address the interconnected factors. We need to develop an appropriate manufacturing strategy to meet market needs and exploit our agility by shaping market opportunities. Agile Manufacturing is therefore based upon the need to adopt a market driven strategic manufacturing response.

With markets becoming more and more dynamic and competition increasingly being based on improved quality and customer responsiveness, the development of an appropriate manufacturing strategy is likely to have a major impact on our survival, success and growth. We therefore suggest that, in general, a strategy for agile manufacturing should be focused on designing an enterprise that will lead to:

- Faster response to highly variable customer demand patterns
- Improved productivity
- Opportunities for system wide innovation, learning and improvement
- Improved product quality
- Better utilization of expensive capital and improved return on investment
- Improved customer and market focus and a better understanding of customer needs and closer customer relationships.
- Flexibility to cope with a wide range of batch sizes including one-of-a-kind production, and a wide range of products.
- Integration of suppliers into product development and manufacturing processes.
- Short-order capability and the ability to rapidly respond to new windows or opportunity
- Capability to undertake multi-venturing through a virtual corporation
- Reduced operating expenses via reduced stock holding, less scrap and rework and reduced system down-time
- Shorter design and manufacturing lead times
- Reduced indirect labour and other overhead costs
- More time and opportunities for management to tackle problems
- System integrity and robustness

These goals provide the basis for the agile manufacturing enterprises strategy. In general, the goals will only be accomplished by investing in organization, people and technology, so the strategy should address those areas. This needs a change management framework for that purpose and involves a different way of dealing with organization and people's issues to that currently used in manufacturing industry. The vast majority of our manufacturing companies adopt a serial engineering approach to deal with these aspects of design. We design and implement manufacturing systems, making trade-offs between cost and performance, and only then deal with organization and people's issues. At this later stage, however, these issues may have turned into problems that have to be dealt with retroactively. It is common to find that we cannot resolve these problems satisfactorily because by this stage many degrees of design freedom have been closed off.

The difficulty here is this: to begin any design, our knowledge about the design problem is quite limited. As the design proceeds, our knowledge grows. However, as we move through the design process, the degrees of design freedom starts to decrease. Thus, when we arrive at the stage of implementation and discover many problems relating to organization and peoples issues, we do not have much design freedom to deal with the problems.

For this reason, agile manufacturing needs a concurrent engineering approach to the design of manufacturing systems and computer-based technologies. This implies that organizational and people's issues should be addressed right from the start of the design process and in parallel with technical issues. This also implies that we adopt an interdisciplinary method, where insights from psychology and organizational science are used to shape the system and address the issues that lie between these various disciplines.

This involves us in making, in addition to the usual cost-performance trade-off, further trade-offs between technical options, costs and performance, user acceptance, motivation, user skill requirement implementation issues, job design and organizational structure. Thus, the benefits of this approach lie in the area of improved design and implementation processes, better system designs, more appropriate organizational structures and motivated and engaged people throughout the company.

The design paradigm for agile manufacturing needs to be based on more interactive design models to enable the early realization of proposed systems using, for example, rapid prototyping and simulation. The spiral life cycle model developed for software engineering (Boehim, 1989), is given for a new design model.

Before addressing the creation of an agile manufacturing enterprise, it is important to understand fully the characteristics and culture of existing manufacturing. Some of the characteristics of our existing traditional manufacturing paradigms will be examined in detail.

4.8 INTERDISCIPLINARY DESIGN METHODOLOGY: ITS IMMENSE IMPORTANCE

Manufacturing system designs is an interdisciplinary activity, and organizational and people's issues need to be addressed in parallel with technical issues, that is, using a concurrent engineering approach rather than a serial engineering one. This means making trade-offs between cost and performance and other factors such as, for example, job satisfaction and user motivation.

One of the primary concepts of agile manufacturing is concerned with the integration of organization, people and technology. To achieve this integration we need to adopt an interdisciplinary design methodology. The design paradigm that should underline agile manufacturing is characterized by:

- A holistic systems based approach
- Concurrent engineering concepts applied to the design of the manufacturing enterprise
- Application of insights from the organizational and psychological sciences to the design of the technology and the overall manufacturing enterprise.
- Consideration of organization and people's issues at all stages, from formulation of business strategy right through to the design and implementation of systems.

Increasing technical sophistication, of course, has been vitally important to the develoment of all aspects of civilization, including manufacturing industry. It could be claimed that technology is the axis, pivot and springboard of development, for without technology there would be no progress. No books to stir the imagination; no cars, no planes, no houses, no radios, civilization cannot develop. If a society has no access to technology, it becomes trapped in a time wrap, that of primitive existence.

Without technology, there would be no manufacturing, but manufacturing is more than technology. It is also about people and how they and technical resources are organized. Manufacturing is about organization, people, technology, management accounting, business strategy, and so on. It is also about the connections between all these dimensions. In the past, we have tended to ignore not only the connections, but also some of the dimensions. We have placed too much faith in technology, using technology to compensate for inadequacies elsewhere, and trying to solve all problems as though they were technical problems.

If agile manufacturing is to be successful, it must break with bad practices. We aim to show how better and more effective manufacturing systems and technologies can be designed based on the insights derived from the relationships between different areas of knowledge. However, to make the transition we need to reform our systems.

5

Concurrent Engineering

Product development teams will be successful to the degree that they practice concurrent engineering principles. Concurrent engineering means much more than sending a manufacturing engineer to the design release review, which, of course, is too late to influence the design as an agile manufacturing.

Multifunctional design teams are the essence of concurrent engineering. The most successful product development teams are the ones with the most complete teams that have all specialists present at the beginning of the project. Much of the success of Japanese product development can be attributed to this, as illustrated by a quote from the $5 million MIT study that compared Japanese 'lean' manufacturers to old paradigm 'mass production' companies.

In the best Japanese lean projects, the number of people involved are highest at the very outset. All the relevant specialists are present, and the project leader's job is to force the group to confront all the difficult trade-offs they will have to make to agree on the project.

By contrast, in many mass-produced design exercises, the number of people involved is very small at the outset but grows to a peak very close to the time of launch, as hundreds or even thousands of extra bodies are brought in to resolve problems that should have been cleared up in the beginning.

Leading-edge product development, derives much of its effectiveness from strong team leadership to maximize the effectiveness of multifunctional design teams. Another MIT study, made in America, regaining the Productive Edge, gave additional insight into Japanese product development project management.

In the Japanese auto companies, each new product is assigned a programme manager who acts as the product champion, carries great authority with the firm, and along with his staff, stays with the product from conception until well past the product launch.

One of the most important early tasks of the product development team leader is to make sure that all issues are raised and resolved. It is so much quicker and less expensive to resolve issues earlier rather than later on, even though the act of resolving these issues earlier may consume certain effort, as pointed out by the same reference.

A key task of the manager is to make sure that all disagreements are aired and resolved at the outset. Achieving consents takes a great deal of effort, but by skillful management at this point it is possible to gain the full commitment of all members of the programme team so that subsequent progress is very rapid.

The disagreements or issues include, but are not limited to, technical feasibility. Other key issues are identifying and then eliminating early, the potential problems in product development, introduction into the factory, production, distribution, use and service. The issues also involve minimizing risk and the effects of changing markets, customers, competitors and regulations.

The manufacturers should take advantage of reducing the cost of variety, rationalizing existing product lines, standardizing parts and processes, making decisions based on total cost, achieving operational flexibility and concurrently designing families of products and processes around modular architectures.

5.1 GAINS FROM QUICKER TIME TO MARKET

Advanced product development coupled with the improved market feedback of mass customization will allow companies to be first to market, thus resulting in the following sales benefits:

- Market availability for product releases will be quicker because new product releases can be based on previous engineering and modules. Further, the improved market feedback from any mass customization offerings will help subsequent products to satisfy customers' needs more quickly and accurately.
- Market share is 100 per cent for the first to market. The earlier a product is introduced before competitors, the longer the 'reign' at 100 per cent. Mass customizers will be able to take full advantage of this opportunity because their flexible operations will be able to quickly respond to rapidly growing demand.
- The first 'mass customizer' gets good, free publicity. Offering a custom product quickly at competitive price is news, especially if it is the first one in a certain market or industry. That, in itself, is news. A newsworthy product introduction has much potential for positive publicity.
- The first to market is perceived to be a leader. Market perception of leadership is of value in generating early sales and commanding price premiums. A steady stream of product improvements or a broad range of mass customized offerings can ensure a continued reputation of leadership.
- The first to market can set standards, either de facto or official. Customers become committed to use a company's products by training, operating system, language, protocol, and so on.
- The first to market can acquire attractive distribution channels. This may be especially important for mass customization, which may depend on good distributors to perform some of the customization at the point of sale.
- Rapid developers take better advantage of windows of opportunity, with respect to both markets and technology. Unless technology is developed internally, competitors are in a race to be the first to develop technology into product. A mass customizer can incorporate new technology into existing platforms using existing modules and previous engineering and software codes.

5.2 TOTAL COST ACCOUNTING

This aspect has been already dealt with at sufficient length in the earlier pages of this document (Section 1.1 & 1.4). To appreciate all the above cost savings and revenue enhancements, it would be advantageous to quantify all costs. If costs were tracked on a total cost basis and in turn the cost saving potential of mass customized products could be known, the products could be given an appropriate and competitive price.

To mass customized products, we should build them to order or excel in niche markets and the products must be built flexibly. This means that one should be able to efficiently and quickly build products that may be different from the ones before or the ones after.

Manufacturing agility is not that hard to accomplish when products are designed right. It is not necessarily expensive to accomplish and does not depend on extensive automation or dedicated information systems. Manufacturing agility can even be achieved in manual assembly. Agile manufacturing capability is one of the key process infrastructures needed for mass customization.

Implementation can be achieved primarily through 'soft' programmes such as Setup Production, Just-in-Time (JIT), Flexible Manufacturing, Continuous Flow Manufacturing, Demand Flow Manufacturing, Build-to-Order, and the various initiatives under the title agility.

The importance of standardization of parts, tooling, fixturing geometries, processes, procedures and raw materials has been discussed earlier (Ch. 2). Standardization is the first prerequisite for manufacturing agility.

Parts must be common enough to be distributed at all points of use without much of setup change for changing the parts. Tools must be common enough to avoid setup delays to change them. Toolings and dies must be designed with standard interfaces so that they can be changed quickly. Fixturing geometries must be common enough to avoid setups related to fixturing workpieces. Processes and procedures must be standardized to avoid setup related changes and learning between successive products. Raw materials must be common enough to avoid the setup of finding different raw materials or the setup for changing the raw materials fed into cut-off saw of screw machines or some other machine tool.

Many standardization efforts can only be fully realized as and when new products and processes are introduced into production. Part standardization generally applies to new products. However, efforts to eliminate duplicates can immediately improve the flexibility of operation. Similarly 'better than' substitutions (1 per cent tolerance parts for 5 per cent parts) can immediately reduce part variety of existing products and possibly improve product quality at the same time.

5.3 SET-UP AND BATCH SIZE REDUCTION

Mass customization strives to produce different products in the same plant with virtually zero changeover time. This requires that the factory must be flexible and agile. If successive products are to be unique and different, there cannot be significant delays to get parts, change dies and fixtures, download programmes, find instructions, or make any kind of manual measurement, adjust settings, or position parts or fixtures. The wasted time of these delays is called setup. Setup makes the factory inflexible and discourages mass customization.

The usual approach to deal with setup is to accept it as a necessary evil and then spread it over as many products as possible in batches or lots to minimize the setup 'change' per part. Industrial engineers have developed formulae for the 'Economic Order Quantity' (EOQ) that determine the 'optimal' batch size based on the setup time. The higher the setup, the higher the batch size and less flexible will be the plant. Efforts are aimed at reducing the setup and thus the batch size. As setup goes to zero, the economic order quantity goes to one, which is true flexible manufacturing.

Lot size reduction is an important step which many leading companies are practicing. Industry Week's **'Best Plant 1994'** Survey of plants manufacture or assemble components in small lot sizes resulted in a median component lot size reduction of 75 per cent.

To achieve this level of agility, which is a prerequisite for mass customization, all forms of setup must be driven to zero. If part variety is too excessive to allow distribution at all points of use, then enough parts for a batch must be assembled into a kit, put together in the raw materials warehouse, delivered to the assembly area, and then distributed to part bins or automation machines. This kitting is a setup that will inhibit flexible manufacturing and mass customization.

The design implications to eliminate kitting are that all parts in a mass customization family must be able to be distributed to all points of use. This should be the focus of concurrent engineering efforts to coordinate the design of product and processes so that this becomes a reality, using standardization techniques.

The product must also be designed to eliminate the need for tooling changes for dies, moulds, tool plates and fixturing or to make the changes as quickly as possible. Parts needing dies for moulding or casting should be designed to be versatile enough.

The degree of success will be largely proportional to how well the products are designed for this environment, which is the overall purpose. If products are designed around too many types of parts to distribute at all points of use, then the parts will have to be kitted and JIT may not be possible. Too many setups in manufacturing can hamper JIT and inventory-reduction programmes. Thus, manufacturing agility is ultimately dependent on product development—the concurrent development of products and flexible processes.

5.4 PRODUCT LINE RATIONALIZATION—EFFECTS ON AGILITY (REF: SECTION 2.1 ALSO)

In most cases, factories can make significant improvement in agility before a new generation of products arrives at. The first step would to be to rationalize the product line using standard techniques. Most likely, products that are eliminated or outsourced are those that have the least 'fit' with the flexible factory. Thus, their removal from operations will probably eliminate most of the oddball (eccentric, non-conformist) parts and the longest setups. In fact, programmes to improve manufacturing agility may be one of the driving forces for product rationalization. The rationalization effort will be even more effective toward this goal if manufacturing agility and its contribution to mass customization and build-to-order is identified as a core competency which is the cornerstone of corporate strategy.

With the oddballs gone or outsourced, the factory can start focusing on improving the flexibility of the products that remain. Since the rationalization effort may have created some major changes in the factory, the next step would be to regroup both the figurative and literal meanings. Plant management should use this opportunity to take a new, objective look at the operation's big picture and question and challenge everything from plant layout to the flow of parts and products. May be the rationalized product line will now permit improvements that were not considered possible before.

Manufacturing management should seek answers to the following questions, about the products lines that remain after rationalization:

- Multi-plant strategies: How does this rationalization affect existing or proposed multi-plant strategies? Can one or more plants be focused on certain groups of compatible products?
- Production line consolidation: Can the new product mix be built on one production line? Or how can they be grouped for the minimum number of lines that would be needed? What would be required to consolidate manufacturing lines?
- Make/buy strategy, what is the optimal make/buy strategy now? Are there opportunities with the 'virtual corporation' or 'virtual organizations' where an agile network, or kei-retrsu, of suppliers delivers parts as part of a Just-in-Time Kanban parts re-supply system? Conversely, are there some critical processes whose outsourcing causes delays that inhibit flexibility? Do vendor/partner relationships need to be developed? There are occasions when some processes must be brought in-house for agility.
- Scope of early improvements. Are there opportunities to focus on a particular group of products or one plant for a 'pilot' implementation?
- Plant layout are machine tools and other product activities arranged by departments or functions. Is there a milling room, a lathe room and an assembly area with parts forced to move in large batches as they work their way from room to room? Does this product line rationalization permit more efficient plant layout possibilities?
- Trends–What trends affect the above issues? Will there be further more ambitious rationalization efforts in the future? What new products are planned and how will their introduction affect operations? Are there any products to be added through mergers or acquisitions?

The scope of any improvement effort is affected by the potential returns. It would help in planning and motivating improvement.

In the mass customization family, for instance, if casting for water meters or gas regulators needs to accommodate various combinations of rest ports and other options, all those features could be moulded or cast into a common casting.

Tools and fixtures can be designed to be versatile enough to accept a common blank, which then can be customized by computer controlled machine tools or robots. The blank must be quickly positioned in the fixture without any need for measuring and manual positioning. Thus, the blank must be designed with common fixturing geometries. Examples of this include the common face-plate blank. Another example involved designing all circuit boards in the same 'form factor' (size and shape) with the same tooling holes so that all of them use the same tooling plate, pallet, or material handling devices.

Single tooling fixtures can be clearly designed to accommodate more than one blank or printed circuit boards. To improve the flexibility of Intel's systems group operations, it was necessary to develop a single tool plate that accommodated seven different circuit board form factors. Pairs of tooling pins were permanently mounted around the central opening of the tool plate. Adjustable support arms could swing out to hold up boards but were kept too small so as not to span the opening. Common tool plates could thus eliminate the setup usually required to change tool plates.

Programmes from computer-controlled production equipment can be changed quickly, with all the programmes stored in a file server that can download the programme almost instantaneously based on identification of the part by some means, such as a bar code.

In manual assembly, even taking time to find assembly instructions or drawings is a setup. To be flexible enough for mass customization, this setup should be eliminated by making the process identical. A small number of choices in assembly could be displayed on a chart in front of the worker. More complex and varied instructions could be changed instantaneously on a computer monitor. If graphical images, say from drawings or illustrations are needed, these can also be displayed on computer monitors, as is done for the assembly of Compaq computers in Houston, Texas. The Apple Computer has even made animated assembly instructions that are displayed on their own Macintosh computers. As with automation programmes, the manual instructions can be changed by input from a bar code wand.

The importance of setup reduction/elimination depends on factors like the volume of production. High-volume mass customization will absolutely require setup elimination and true agile manufacturing. Low-volume production may tolerate somewhat greater setups, but the cumulative cost and time of such setups must be considered. The goal should be to make mass customized production of a variety of products as efficiently as the mass production of one product.

Inventory reduction just as standardization as pointed out in earlier chapters (Ch. 2) is the prerequisite to setup reduction. This setup reduction is the prerequisite to inventory reduction and just-in-time programmes. Like many innovative programmes, the just-in-time label generates its share of misunderstanding. Critics and many people who do not understand the concept think that just-in-time only consists of delivery trucks skidding up to loading docks with parts delivered 'just in the nick of time' before production halts. Further, they also think that incoming inventory has been simply shifted from the assembly plant floor to the suppliers' warehouses. The perception is sometimes true when suppliers want to do business with Just-In-Time customers but are not themselves, 'with the programme'.

However, the real goals of Just-in-Time programmes are inventory reduction, flexibility improvement and continuous improvement forced by the elimination of 'Just-in-Case' inventory. It is the flexibility improvement that is of interest to companies implementing agile paradigms like mass customization, build-to-order, and niche markets. But being able to build customized products in 'lots of one' requires the virtual elimination of setup and inventories rather than just their reduction.

America's leading companies implement JIT as part of a general improvement strategy. Industry Week's 1994 'Best Plant' survey of the 25 top companies showed that 96 per cent of these companies had JIT/continuous-flow production. However, many companies have had varying degrees of success attempting or implementing JIT/inventory reduction, and so on.

Similarly, a build-to-order system 'pulls' parts from suppliers on a just-in-time basis, thus eliminating parts inventory along the supply chain. Regardless of who 'pays' for this inventory, the customer ultimately must pay a higher price. Eliminating excessive supply chain inventory costs will allow customers to pay less for equivalent products.

Companies known for rapid deliveries, like Federal Express, provide companies with 'inventory-less' direct deliveries of parts and products to and from factories. After National Semiconductor commissioned Federal Express to run national's storage, sorting, and shipping activities, delivery time was reduced from 45 days with an ultimate goal of 72 hours. At the same time, distribution costs were also reduced from 2.6 per cent of revenue to 1.9 per cent. Adding the value of increased sales from customer satisfaction point of view would make inventory-less distribution even more attractive.

Less interest expense will be incurred for expensive components in products that sell sooner because of increased throughput of flexible plants and limitation of finished goods inventory.

Inventory write-offs can be eliminated because there would be no products in inventory that could deteriorate or incur damage or become obsolete. If there is a substantial finished goods inventory of an expensive product at the end of that particular product's life, the obsolescence write-off can be enormous.

Realistic early production feedback can be obtained from 'pilot' manufacturing runs in a real (flexible) manufacturing environment rather than in separate pilot manufacturing runs in a real mass production plants which generally do not have the capability nor the patience to experiment with pilot builds. So pilot runs are built in pilot plants with special machinery (not production machinery) by prototype technicians who exercise more skill and perseverance than typical production line workers. Because of this, pilot runs in pilot plants can be successful, but the product can still have a difficult launch in real production environment. The result of realistic feedback at the pilot stage is, lesser introduction cost and quicker product introduction ramps.

In addition to the above cost savings, mass customization can also improve profits through increased revenue.

Customers' needs are satisfied better with customized products as these products are exactly what they want. The increase in sales would be the difference between customers that would be satisfied with standard products and customers that would be satisfied with customized products.

Price premiums will be possible for customized products, since they meet customers' needs better, and should be worth more to customers. The customized product, in the form the customer wants, may have little or no competition and, thus, may enjoy somewhat of a monopoly with all the ensuing pricing advantages. Price premiums often range from 10 per cent to 50 per cent.

Mass customization is a paradigm shift that will forever change customers' expectations. Today, customers may be reluctantly buying the product that comes 'closest' to meeting their needs at an attractive price but they will no longer be content with 'close'. They will now expect exactly what they want and that is worth something to the customer.

Mass customizers will have the pleasant decision of whether to apply cost savings and premium pricing opportunities to the bottom line or pass these savings on to the customers to compete better on price and expand market share.

Broader market can be served by versatile designs and flexible plants.

Mass customizer enjoys an ever greater opportunity. One of the effective cost-reduction tools of mass customizers is to eliminate most of the distribution chain as we know it, being able to build to order and ship from the plant eliminating warehousing and associated distribution costs from the plant to the customer.

5.5 LIFE CYCLE COSTS

An often neglected, part of total costs are life cycle costs, which are those costs that are incurred over time, such as service, repair, maintenance, field failures, warrantee claims, legal liabilities, changes over the life of the product, and subsequent product developments. Products can be designed to minimize life cycle costs. The cost of changes can be minimized by a methodical product definition and thorough product development. Change costs and subsequent product development costs can be minimized with modular product architecture, where many modules can remain unchanged even as products are updated or designed.

One clear result of the success of Japanese companies and others that embraced the continuous improvement model was the tremendous force for change. Company after company and industry after industry had to increase their quality and lower their costs to compete in the face of a clearly superior business model to mass production. But a second result, perhaps less clear but of equal importance, was that all of this improvement taught customers to become more demanding. Once they discovered they could get both low costs and high quality from the company, they stopped putting up with poor service from others and began demanding that companies provide goods and services exactly as promised. Indeed, these two effects were reinforcing, driving each other to greater and greater levels, until in most industries continuous improvement only got them into the game. Quality was no longer a differentiator.

We are nowhere near the point yet with customization, but companies in the next decade would be faced with the same precise predicament. Once customers begin to see that they can get products made just for them that almost exactly match their needs at a price they are willing to pay, you can believe that they are going to ask for it elsewhere. Mass customization is indeed moving out of the frontier and becoming an imperative.

5.6 CONCURRENT ENGINEERING THROUGH TEAMWORK (REF: ALSO SECTION 5)

A large number of manufacturing organizations are compressing new product development lead times in order to remain competitive. To achieve this, they are practising concurrent engineering through the use of teamwork. This is achieved by harnessing their people through flexible multidisciplined teamwork. Though, the business processes are well supported by computers but difficulties with mobilizing and managing teams still challenge the progress. There are three main issues related to teamwork in the manufacturing process: firstly, the difficulties in identifying the characteristics of efficient and effective teamwork and their effects on the manufacturing process. Secondly, various approaches for analyzing teamwork and difficulties of obtaining information on team performance problems. Thirdly, the way in which current information technology supports the teamwork and the sort of tools that are needed to facilitate the communication among team members to enhance work efficiency and quality.

5.6.1 Characteristics of Teamwork

Manufacturing is both a technical and a social activity performed by a workforce who continually exchanges information. Typically, people involved in the manufacturing process are part of a multidisciplined team, working both individually and collectively, and trying to apply creative problem solving techniques in order to translate customer requirements into a technical product. Each member has a position and an expected role to play within the team, and probably comes from different cultural and education backgrounds, possesses different personalities, and has developed his/her own perception through personal experience of what is considered teamwork. The main objective of teamwork is for members to be able to communicate intentions, ideas, and information to each other to maximize the work performance of the team. The approach and style that is used to carry out such a task will be a function of the organizational goals, the types of team members and their interaction and the qualities of the team leader.

5.6.2 Concept of Teamwork

It is important to first identify the different types of activities performed by individuals and by groups. For a lot of activities associated with the problem solving process, there is no apparent difference. In case of simple problems, an individual is capable as a group producing a working solution. If the group is to realize its full potential and be able to tackle complex problems it will need to work effectively as a team and therefore members must communicate with each other. It appears that it is very difficult to distinguish the boundaries between individual work and teamwork, because most individual activities must be communicated and used by other team members. Perhaps it is better to distinguish between work and activities where work encompasses both activities and communication process of exchanging information. Teamwork must also be defined in terms of its context because it will be different from one situation to another. For example, a distinction must be made between working together as a large all encompassing group (possibly the whole company) and working together in much smaller groups, say two to six people. A person may be a good team player in one situation and feel very uncomfortable in another situation. Furthermore, there seems to be a need to identify the type of activities which should be performed individually and together, to determine the optimum balance and to determine the most suitable conditions which promote optimum work efficiency.

5.6.3 Barriers to Teamwork

In a typical team, there is a mixture of personalities, educational and cultural backgrounds, different vocabularies, priorities, and motives among team members. Considering these differences, a major problem which members of a team commonly face is the intellectually threatening situation which damages the confidence of the individual and the morale and enthusiasm of the team. Emphasis must be placed upon the reasons why such a situation occurs, how it can be avoided and how it should be handled.

5.6.4 Personality and Language Barriers

Dominating personalities have a tendency to create these threatening situations and also control the teamwork process because it is they who are taking the initiatives. In this case, problems can readily arise because these individuals do not have the awareness, understanding, perceptiveness or consideration of other people's personalities, desires or motivation. Consequently, dissension among team members is created due to the lack of respect which is conveyed via verbal comments, facial gestures or by actions (or lack of actions) taken by an individual. The use of unfamiliar vocabulary has the tendency to amplify intellectually threatening situations, because the members do not understand each other and do not communicate effectively. This is very typical with multidisciplined situations in which case it becomes the responsibility of the team leader who is also the facilitator of such problems, to disseminate a clear idea of teamwork expectations to the team members.

5.6.5 Educational and Status Barriers

A problem of concern arises while attempting to include less formally educated people (but perhaps more experienced) in the team and how to capture their knowledge. For example, a young highly educated engineering graduate starts employment with a company with responsibility of managing a technician. The engineer has a perception that he is expected to know more than the less formally educated person. However, the engineer does not realize that the practical experience of a time-served technician is equivalent to or even greater than a formal academic education which is highly theoretical. This egoistical attitude many times results in the technician feeling unimportant and losing respect for the management. The Japanese approach for dealing with such problems is quite different. For instance, in Japan, the managers and graduate engineers are expected to perform tasks that are normally done by employees who are lower in the organizational structure hierarchy. An extreme example of this is the sweeping of floors on the production line. This system has evolved partially because the companies send their managers and graduate engineers to the production line for training, partially because of the Japanese culture where there is an emphasis on the adherence to conformity. Obviously, the approach will not work in every culture, although some UK companies have adopted similar schemes.

5.6.6 Organizational Barriers

The corporate culture and policies within an organization are critical to the shaping and evolution of the team. The manner in which culture is promoted can make or break a team. There must be a mutual understanding between the team members concerning each other's expected roles, goals and priorities, technical needs, training and education, rewards and incentives. This must also be realized between the team members and the team leader, in order to maintain the team's compatibility with the corporate aims of the company. Expectation and perception of roles can lead to problems within a team and can be very difficult to clarify. In a conventional hierarchical organizational structure, the details of responsibilities and duties are defined between a manager and an employee on an individual basis. In a team-based setup, there is a tendency for the roles of team members to be obscure, with team members only having a superficial knowledge of each other's duties. This allows for a large degree of freedom for misinterpreting the boundaries of roles and responsibility. Furthermore, roles are generally not defined because they encompass social aspects, they evolve with time, and they are continuously being renegotiated. In addition, the way in which a person perceives his/her own role or someone else's role, may be entirely different from the way someone else perceives it.

5.6.7 Lack of Common Motives

An insufficient understanding of each other's goals and motives, a lack of trust in each other, and job insecurity have a detrimental effect on teamwork and productivity. As a consequence, team members may compete aggressively, limit co-operation with each other, and hold back on information or ideas. Competition can be healthy, but when it is taken to its extremes, the consequences can be very damaging to the team. Unfortunately, the building of trust can only happen over time and often a single incident, however trivial it may be can signal a lack of commitment, make it difficult to rectify the situation and restore it once again. Job insecurity is also quite common among employees especially if the company is going through some difficult times or is affected by a recession. By hiding information which other people do not know, an employee can become more or less indispensable since no one else within the company can do that job. To a large extent, most of these problems are unavoidable; however the organization should try to minimize their severeness through effective management.

5.6.8 Training and Education

The training and education of team members are critical both in terms of having the proper mix of technical expertise and having the necessary knowledge and experience regarding teamwork. Selecting the proper combination of technical expertise is rather straightforward, but finding the appropriate combination of personalities of team members can be a difficult task. One of the major problems in achieving this is that most educational systems do not actively teach or promote working in teams. Thus, teamwork characteristics emerge out of the personality traits and evolve over time as one gains experience. A reason for this is that in general, engineering curricula are dominated by analysis and are focused at rating individual performance in solving problems. A critical component to the evolution of teamwork and enhanced efficiency and effectiveness is that every individual within the team has a natural need for a feeling of importance and that each has made a key contribution to the development of the product. This can be actively promoted through recognition of achievement, healthy competition, motivation and incentives and a good balance between team and individual accomplishment.

5.6.9 Teamwork Analysis

Understanding the processes and the distinguishing factors that affect teamwork will pave the way for reducing the problems and barriers affecting the performance of the team. The process of teamwork can be generally categorized into a reasoning process and a social process, and analysis techniques should attempt to separate these two aspects. The analysis can be either formal or informal, and occur on a regular or irregular basis. Either way, the feedback generated will be useful for regulating team behaviour and hopefully raising team performance. A major concern with teamwork analysis by observational protocol studies and peer reviews is how subjectivity should be treated. The use of a neutral observer in teams can reduce the degree of subjectivity. However, completely removing subjectivity is impossible and should be discouraged. It is far better to learn to manage it by introducing it in different ways and actually obtaining additional information in the process. For instance, the social situation can be changed to capture various types of teamwork activities. It is also important that observations should not be rated in any way. Another problem that often occurs in teamwork analysis is the capturing and selection of relevant information (quality and quantity). Capturing of too much information can be equally bad as that of capturing too little information. Perhaps the use of handwritten notes rather than utilizing audio or visual recording media would better serve the capturing of relevant information and reduce the quantity to a manageable amount. The filtering and recording of information would then be done simultaneously but be limited to the ability of the observer.

5.6.10 The Way Forward

The benefits of concurrent engineering are clear, but the application of teamwork principles to different business situations is more problematic. The primary objective should always be to work together and to achieve best engineering practices, through total team commitment and effective team leadership. When this is established and is working satisfactorily, then the use of supporting tools should be considered to enhance the process.

6

Ideal Production, Synchronized Manufacturing

The ideal production plant will set the production capacity for each work cell and each operation is synchronized with operations that come before and after. Unfortunately, there are no ideal production plants. However, JIT manufacturers strive towards the goal of totally synchronized production in which each work cell receives exactly what it needs and when it needs.

Several problems result from lack of operation in synchronization. Large WIP inventories are possible even when batch sizes are small and setup times are short. Shop-floor queues are usually the result of bottleneck operations, poor production planning and part shortages.

Bottleneck (or limiting) operations are those with less capacity and large loads. These operations may be considered the weak link in the chain of production processes that go into making the finished product. For example, bicycle manufacturers might be able to make 5,000 wheels per day, but can only make 2,000 frames. If the manufacturers continue to make 5,000 wheels each day, WIP inventories will build up because another stage in the process can only cope up with 4,000 wheels (assuming there is no requirement for spare wheels).

Better synchronization can be achieved by approaching the problem from two different perspectives; production planning and shop-floor execution.

- Production planning: Can be planned so that there is near perfect synchronization between loads on each work cell. Such perfection can be achieved by careful master scheduling and detailed capacity planning through limiting work centres. Some available software attempts to optimize production schedules by automatic optimization of bottleneck work centres. This software is limited in its applicability and has yet to find general acceptance.
- Shop-floor Execution: Most JIT manufacturers divorce production planning from shop-floor execution and use some kind of 'intensity pull' method on the shop floor. A shop-floor pull system merely means that each supplying work centre does not make anything until the next work centre issues Kanban cards to the supplying work centre. This approach can yield excellent synchronization, flexibility, and minimum inventories, provided each work centre has a short setup time, small lot sizes and short cycle times.

The closer a production plant comes to synchronized production control, the nearer it will be to just-in-time manufacturing. Such an achievement requires excellence in every area of the production.

6.1 VENDOR RELATIONSHIPS

A traditional manufacturer tends to have an adversarial relationship with vendors. Procurement personnel are judged primarily on purchase price variance; in other words, they are rewarded for getting the lowest price. The company will ensure that there are several suppliers for each part so that these vendors can be pitted against each other when contracts are negotiated to ensure continuity of supply.

In contrast, agile manufacturers employ mutually beneficial vendor relationships. Vendors are selected primarily on the basis of quality, delivery, reliability and flexibility. The theory is that the benefits of reliability, consistent quality and just-in-time deliveries are more valuable than a few pennies off the price.

Techniques required for a good vendor relationship include single sourcing, certification, and openness. Single sourcing uses just one supplier for any component or raw material that is purchased and thereby reduces the total number of vendors. The idea is simple: a supplier doing a lot of business with a company will treat it with more care. When large orders are going to fewer suppliers, then the company will be able to create closer, more cooperative relationships with each supplier. This approach has been used very successfully by companies such as Xerox, which reduced the number of vendors from 5,000 to below 200 as a part of its successful JIT quality programme.

Plunging into a single-sourcing arrangement will be irresponsible if vendor performance records are not known. Performance is certified by a process that assesses every supplier record of quality, delivery, reliability and flexibility. Deliveries from a certified vendor will be received without the need for incoming inspection and will often be delivered directly to the shop floor.

Vendors are certified to supply particular components when performance in each aspect is up to standard. If a vendor does not match company standards, then the buyer will seek to resolve the problems with the suppliers. As the certification process develops, many vendors will be dropped out and the total number will be reduced. Eventually the company will purchase components from only a small number of certified suppliers.

Unfortunately, vendor certification recently has gained a negative connotation because some large companies have intimated smaller suppliers into providing concessions on service and price. Many large companies have used their clouds to force smaller vendors to hold large inventories of finished components in order to support their customers' just-in-time requirement. This may only push wasteful safety stocks back onto the vendor rather than allowing the vendor to move into a world-class arena. This approach is contrary to agile vendor relationships, where mutually beneficial long-term relationships are the primary aim.

A good starting point for improved relationships is openness. Agile manufacturers have an open door policy with vendors. There is free exchange of technical information, clear communication of needs, expectations and performance and the sharing of financial information about products being supplied. This open communication provides a basis upon which trust can be built. As these relationships progress, further contribution results in better designs, higher quality and lower costs.

Agile manufacturing needs unprecedented levels of quality and reliability, just-in-time deliveries and short lead times and often requires additional services such as deliveries in sequence or special point-of-use packaging. In return, the vendor receives large orders, long-term contracts and fair prices. A vendor will often require just-in-time payment for just-in-time deliveries. Thus, a happy circle frequently develops as suppliers themselves move into agile manufacturing. The techniques required to provide the superior levels of quality and service have the effect of making the supplier much more productive and product price can be reduced. In fact, many companies are paying less for components and raw materials while receiving JIT deliveries and zero defects.

6.2 WORKFORCE MANAGEMENT

Frequently, the most difficult problems to solve when introducing agile techniques relate to the management of workforce. A company committed to agile manufacturing must be prepared for fundamental change in management style and philosophy. After years of mutual distrust and poor relationships, both managers and production personnel often find it difficult to make the required changes.

Many plants have discovered that shop-floor operators, appropriately managed, can make the largest contributions to the attainment of agile goals. Technological changes and automation have taken much of the physical workout of the production process. Agile manufacturers attempt to make full use for their workforce hands-on knowledge to make products better, faster and safer.

New approaches to the management of workforce can take several months or even years to introduce. What is required is trust, honesty and success. Developing trusting relationships between people and managers takes time and experience.

6.3 TRANSFER OF RESPONSIBILITY

A traditional manufacturer tends to treat production personnel as only suppliers of labour. Though, middle managers and supervisors are employed to make day-to-day decisions, resolve problems and take responsibility for production, an agile manufacturer will take great steps to give operators greater control of their daily work. They will have prime responsibility for product quality, for scheduling preventive maintenance and for attaining production targets. This change does not mean that middle managers, engineers, and specialists are no longer required; however, their roles change.

6.4 EDUCATION AND CROSS-TRAINING

Agile manufacturers dedicate a great deal of time and money to educate employees. Some of their efforts is devoted to ongoing training in JIT manufacturing, quality control and customer service. Another important area is the training of operators to do a wide range of tasks within the production plant.

This cross-training of people allows manufacturers to have greater flexibility because operators can be moved between tasks. If customer demand is high within one product group and low in another, then people can be moved to an area where they are not needed. If product mix is such that the production rate on different work centres has to change so that synchronization is maintained, then the number of people in each work cell can be changed.

6.5 PROBLEM SOLVING AND QUALITY CIRCLES

The introduction of quality circles has been received with mixed opinions in the West. The purpose of quality-circle approach is to have every employee involved in solving production and other problems. No matter how this participation is, the goal is to have the active participation of all employees, each of whom can contribute his or her own individual skills and experience.

In the past, problems were solved by middle a managers, engineers and specialists. In an agile plant, the entire workforce is involved in one or more projects aimed to continually improve products, processes and services. These programmes have found spectacular success in many companies because an atmosphere of teamwork and common cause have enabled people who previously had very little opportunity to contribute to become innovative and resourceful problem solvers. In addition, the people involved in these efforts enjoy their work because they have a wide variety of tastes and their ideas are treated with respect.

6.6 FLEXIBILITY/AGILITY

While Western manufacturers are working hard to attain world-class status in terms of quality, delivery, performance and customer service, many top Japanese companies are building upon existing excellence. A major challenge facing Western manufacturers in the near future will be "flexibility".

At one time, the belief was that techniques of just-in-time manufacturing were suited only to repetitive production processes with highly predictable schedules. While it is true that JIT is easier in such an environment, many companies are now extending just-in-time concepts to more complicated production situations where product quantities and production mix change frequently.

In their 1988 report **"Flexibility: The Next Competitive Battle," Professors Jeffrey G. Miller, Jinchiro Nakane, and Around De Meyer** showed that many leading Japanese manufacturers emphasized on production flexibility and price as their future competitive edge. They strived to make products at low cost and with a high degree of flexibility to meet customer needs.

Two aspects of flexibility are important: production flexibility and design flexibility.

- Production flexibility is achieved when the company can offer short lead times, when product mix can be changed significantly from day to day, and when people are cross-trained to manufacture a wider variety of products. A company that can offer this level of flexibility (without a price penalty) has a significant competitive advantage.
- Design flexibility is related to the company's ability to introduce new products and modifications to existing products. Market needs changes quickly. A company must understand current and future customer needs, develop innovative products and deliver those products to marketplace quickly. In the early days, seven to ten years were needed to bring an automobile from conception to production. Today, Nissan can introduce an entirely new car in less than three years. Recently, Chrysler has been able to achieve similar levels of new product introduction. This level of flexibility to meet customer needs in both short term and long term is the mark of a company truly committed to agile manufacturing.

7
Agile Manufacturing: A Bird's-Eye View

7.1 INTRODUCTION

The international business scenario is undergoing a turbulent change. The time-honoured strategy of mass production/economies of scale has become irrelevant. The reason being the requirements of low volume, high quality, customized and specific products. These products typically have short life cycles; shorter lead times and are information rich. Interestingly, additional services and value additions (like product upgrades, future reconfigurations, etc.) have become as important as the product itself. This paradigm shift is summarized in Table 7.1. In short, everything is changing fast and unpredictably, paving way for the new philosophy of **agile manufacturing** or **agility**.

TABLE 7.1: Manufacturing paradigm shift

Old	New
Product-driven	Customer-driven
High product variety is bad	High product variety is good
Product replacement	Product upgrading
Functional orientation	Process orientation
Economies of scale	Economies of time
Dedicated equipment and jobs	Agile equipment and jobs
Single enterprise	Network enterprises
Information dynasties	Open "seamless" information
Top-down leadership	Vision-driven

Agility is the ability to thrive and prosper in an environment of constant and unpredictable change. It is a business-wide capability that embraces organizational structures, information systems, logistics, processes and in particular, mindsets.

A more straightforward definition of agile manufacturing is provided by David Caruso of Visibility Inc. as "Agile manufacturing is a flexible manufacturing model that enables manufacturers to build and deliver a wider mix of customized products, fast and cost effectively".

Agile manufacturing (AM) is distinct from lean/world-class manufacturing. Though both aim at a common goal (greater success, profitability), the principles adopted are different. Leanness is concerned with eliminating 'wastes' while AM looks at exploiting opportunities through harnessing knowledge, being flexible and nimble. Cost is the market winner for lean manufacturing while it is service levels and responsiveness for AM. AM recognizes the volatility of change and puts mechanisms in place to deal with it.

The four steps to agile manufacturing are shown in Table 7.2.

TABLE 7.2: The four steps to agile manufacturing

Traditional Manufacturing	Gaining Control	World-Class Manufacturing	Agile Manufacturing
* Complex systems	* ERP or MRP II	* Lean Manufacturing	* Enriching the customer
* Departmentalism	* Better customer	* Just-in-Time	* Competitiveness
* EOQ	service	* Total Quality Mgmt	through cooperation
* No employee involvement	* Reduced inventory	* Much less costly	* Organizing for change
* Financial secrecy	(10-25%)	* More responsive	uncertainty
* High inventory	* Lower production	* Long-term	* People and information
* Inspection	costs	profitability	* High customer flexibility
* Lack of strategy	* Greater flexibility	* LT improvement	* Integrated flexibility
* Late delivery * Long cycle times	* Better control * Planned operations	* Productivity improvement	* Technology * Highly educated &
* Record inaccuracy	* Better	* Time to market	trained workforce
* Shortages/Expedite	communication	* Zero defects	* Flexible management
* Month-end push		* Inventory turns	* Virtual cooperations

At the core of AM are the four cardinal principles with their slogans:

(i) Customer Prosperity — " Enrich the Customer"
(ii) Fitness for Change — "Master Change and Uncertainty"
(iii) Leveraging Resources — "Workforce and Information"
(iv) Virtual Organizations — "Cooperate to Compete"

These can be summarized as in Table 7.3.

TABLE 7.3: Principles of agility

Agile Principles	*Enrich the Customer*	*Master change & Uncertainty*	*Workforce and Information*	*Co-operate to Compete*
Top-level Agile Strategies	Solutions are :- - High value - Mass customized; niche - High info. content - Value priced - Multiple	- Change proficiency - Reconfigurability operations - Rapid response	- Entrepreneurial environment - Exploit all skill and knowledge	- Choose to partner - Virtual organizations

Contd.

Principal Agile Capabilities	- Continuous market analysis - Intense customer focus - Customer-driven quality - Networks	- Process re-engineering - Intense customer focus - Customer-driven quality - Networks	- Continuous communication - Teaming and trust - Open learning value-based compensation - Networks	- Partnering codes - New legal style - IP management - Virtual work skills - Networks

7.2 CUSTOMER PROSPERITY–"ENRICH THE CUSTOMER"

Agile manufacturing is responsive to the ever-changing needs of the customer, but at the same time realizes that customer won't necessarily pay a premium for quality. Firms must have a thorough understanding of their customers and supply them with products and services that add value to their lives. It requires a short, medium and long-term views. It requires companies to understand their customer's use of their products more thoroughly than they know themselves. To address the customers' real needs they must sell solutions and not products. They may have to add extra services or technical support or special terms. They may even need to add complementary products supplied by other companies–perhaps by their competitors–to truly satisfy the customers needs. An integration of the design and manufacturing processes becomes mandatory implying the adoption of concurrent engineering and CAD/CAM approaches.

Design and development of a product can be enhanced by a direct participation of the customer. Attention should be focused on configurability, modularity & upgradeability of products and long-term satisfaction of customers. Such cooperative endeavours will lead to development of service-rich products that can evolve over time. A direct benefit is long-term customer relationship.

Agility implementation widely employs ISO 9000, TQM and BPR as 'means', but roots them in the 'ends' of a company wide strategy of sustained competitiveness as shown in Table 7.4.

TABLE 7.4: Agility compared to other standards

	ISO 9000	*TQM*	*BPR*	*Agility*
What	Prescriptive Standards for QA	Philosophy and strategy for product and service quality	Strategy to re-organize work around processes not tasks	Enterprise-level direction, strategies and business models
Why	Assure Predictable quality to enhance trade	Surpass customers' expectations of quality	De-fragment and re-align business processes	Sustain competitiveness in the face of change
How	Audit of processes	Adoption of continuous improvement practices	Interventions to redesign specific business processes	Adoption of agile strategies and best practices
Metric	ISO 9000 Certification	Baldrige and other awards	Internal citation of benefit	Profit and imitation
Benefit	Customer Confidence	Satisfied customers	Internal efficiency	Global competitiveness

EXAMPLE: The great range, options and degree of customization offered by **Dell Computers** have made it the current market leader.

Honda's new range of bikes comes with an electronic key containing information that changes the performance of the machines according to the rider's choice.

7.3 FITNESS FOR CHANGE — "MASTER CHANGE AND UNCERTAINTY"

The challenge before organizations is to survive first and take advantage of the turbulent environments. This implies that organizations need to be flexible and nimble to accommodate changes. This mandates not only the use of flexible equipment, but also flexibility in the organizational structure itself. An agile organization delegates considerable authority to its employees who directly interact with the customers. This is essential since these frontline people are the sensors of the company and will be the first to recognize changes. They must be given local decision-making authority to act on these changes.

An agile organization will often need more than one organizational structure at the same time. Leadership, motivation and trust must replace traditional management style of command and control.

Exhibition of better and quicker responsiveness to customers often requires physical proximity and excellent communications. To meet this challenge, agile organizations will have smaller production and service centres geographically spread out.

EXAMPLE: The worldwide network of **Dell Computers** and their extended organization.

7.4 LEVERAGING RESOURCES–"WORKFORCE AND INFORMATION"

The skills and knowledge of workforce are the most important assets of an organization. Knowledge includes product knowledge, experience as well as a deep understanding of customer's needs, anxieties and service requirements. An agile company invests substantially in training and developing its human resources.

Availability of complete, pertinent and easy to use information is becoming a key tool for ensuring customer satisfaction. The best way to create customer awareness is to provide a great deal of information to both customers as well as to people within the company (product information, training, specifications, upgrades, etc).

Maximum utilization of Information Technology and Knowledge Management tools is essential for agile organizations. Open sharing of information is a key aspect of agile operation.

EXAMPLE: **IBM's worldwide information system** linking their 350 partner companies provides technical information, order statuses, communication, helpdesk facilities, etc.

7.5 VIRTUAL ORGANIZATIONS–" COOPERATE TO COMPETE"

No company can have all the required skills, knowledge and resources to meet all the market needs. In order to meet these requirements, opportunistic alliances of core competencies across several firms (even competitors) is a strategy adopted by agile organizations. These are known as virtual organizations. These are not traditional joint ventures or mergers, but informal, cooperative arrangements without complex legal structures. Everyone understands his role and expectations. To be successful, these require considerable trust, respect and openness to try unorthodox approaches. Information technology is a key enabler in these structures.

EXAMPLE: Cooperative venture between HP and Intel to develop the 64-bit "Itanium" processor.

7.6 USER-FRIENDLY TECHNOLOGY

The new market standards would not have been possible without an equally new role for technology. In traditional manufacturing, for instance, machinery was hard wired for narrow purposes. Each machine produced a piece of the product. A new product or a new piece required a new machine. The fragmented markets faced by the Japanese and Europeans, however, required a more flexible use of existing technologies. In the 1950s, the Europeans and Japanese developed work processes for using narrow purpose and relatively inflexible equipment more fluidly. For instance, they learned to use team-based production to reset machines or roll different machines in and out of the assembly line quickly to reduce downtime when changing from one version of a product to the next.

Eventually, information-based technologies allowed employers to locate flexibility in the technology itself as well as in the work processes for using the technology. With flexible software, a few keystrokes at a control board can reprogramme the machinery and work processes. The computer has brought a whole new level of built-in flexibility and precision in production and service delivery. Those who have exploited fully the flexible new technology by using it in conjunction with equally flexible workforce and organizational formats, have raised the level of competition and increased the range of competitive standards. For example, the computer's precision allows employers to raise productivity and quality standards. The computer saves time by allowing workers to make changes with a few keystrokes rather than reconfiguring whole production systems on work processes. The new technology also provides convenience. User-friendly software makes technical complexity as invisible to most customers and workers as the carburettor is to most drivers of cars.

7.7 ROBUST PRODUCTIVITY

Productivity is pursued differently in the new economy as compared with the old. In the organizations of the old economy, white-collar and technical staff increased productivity principally by rationalizing organizations, mechanizing work processes and reducing personnel costs by using fewer or cheaper employees. The essential goal of the productivity strategy was greater efficiency–more output for less cost. The main target for cost cutting was personnel costs, because they represented a lion's share of costs in every organization. By automating work processes and instituting rigorous organizations design, employers in the old economy used cheaper labour by reducing skill requirements and realized even more substantial savings by reducing the size of the workforce.

The old-time religion of productivity with a single end which focused on cost reduction does not work in the new economy. Highly rationalized bureaucracies are too rigid to respond to the fast pace of change that characterizes the new competitive environment, namely, the organization of the economy board or authority and its resources at the top. The presumption is that general access to authority and resources will result in profligacy and waste. Yet, in the new economy, access to authority and resources is required at the point of production and service delivery and at the interface with the customer if the organization is to provide quality, variety, customization, convenience and timely innovations. Moreover, authority and access to resources are required down the line in order to encourage full utilization of the new flexible information and communication technologies at the core of the new economy.

Increasing productivity by reducing costs results in lean organizations, narrow purpose technologies and unskilled workforces that are cheap but too inflexible and anaemic to respond to the new, broader set of competitive requirements. The new economy requires management, technologies and workforce that are flexible and robust. In the old economy, organization, technologies and workforces were the targets for cost reduction. In the new economy, they are the resources to be developed in order to add value.

7.8 SUSTAINABLE CUSTOMIZED LOW WASTE GREEN PRODUCTS – AGILE MANUFACTURING HOLDS THE KEY

Meaning of Sustainable Agile Manufacturing: According to David Caruso, Vice-President, Visibility Inc., Wilmington, Mass., USA, manufacturer of software's for agile manufacturing, agile manufacturing is a more flexible manufacturing model than the flexible manufacturing itself that enables sustainability to manufacturers to build and deliver a wider mix of customized products fast and cost effectively.

Agile Manufacturing Approach (AMA): The goal of AMA is to discover and codify guiding principles for manufacturing, educate future leaders for the manufacturing workplace and otherwise infuse important principles and technologies into manufacturing practice.

Manufacturing has to be viewed as a broad-based activity ranging from product design through production, product use, maintenance, prevention and reuse of wastes, etc. But AMA recognizes as an interdisciplinary activity that it requires seamless integration of technology, management, information system, workforce, etc. It approaches these aspects holistically rather than take rifle-shots at individual issues, to achieve quick fixes by bridging the traditional technology, management, workforce, etc. with a broad understanding of manufacturing that integrates key functions and disciplines involved in creating, making, selling/servicing products, etc. It encompasses not only critical operations within the corporations such as product, process, engineering, administration and marketing/sales/services, but also vendors/suppliers, customers, community, government, etc., outside the corporation as shown in Fig. 7.1.

Thus, agile manufacturing enables a manufacturer to switch over fast from one product that is being produced to another product desired to be produced in a cost effective manner within the framework of the system. Agile manufacturing includes the entire business processes, design, tooling, machines and machinery layout, materials and inventory, waste prevention and elimination, cost specifications, price constraints, marketing and service and support, quoting and delivery dates, etc.

Added to these, achievement of supply chain excellence would enable manufacturing companies to respond to business and consumer imperatives and the key challenges ahead of them.

This part of the article briefly attempts to look at some pitfalls in contemporary strategic thinking, the business imperatives for the current millennium, how supply chain excellence would enable manufacturing companies to respond to these imperatives, where is Indian business today with respect to supply chain excellence and what are the key challenges ahead of them.

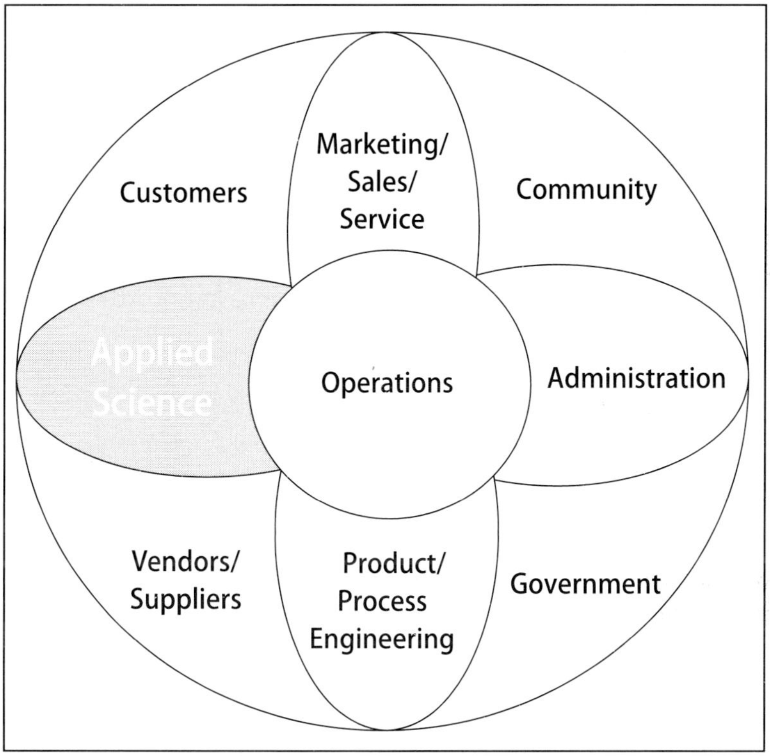

FIGURE 7.1: Agile manufacturing approach (AMA).

Supply Chain Management

During the last decade, a large number of companies worldwide have invested large sums of money and years of effort to improve, streamline, reengineer and reconfigure their supply chains since supply chain performance impacts the top line (increased margins due to reduced cost and improved asset management). However, the returns on these have not been to the desired levels. The key reasons for this are:

- Most improvement efforts to date have emphasised tactical efficiency in one piece/segment of the supply chain.
- Organizations have often undertaken long strategic reviews creating execution plans.

- The required human and organizational performance have not been factored into the execution plans or have been considered too late in the change process.

As a result initiatives that began with great intentions of enhancing competitive advantage have often been reduced to struggles to comply with customer demands, financial objectives and/or regulatory requirements.

Taking an end-to-end view of the supply chain, leading organizations are continuously rethinking the roles and capabilities that are performed with the company and those performed by the trading partners (suppliers, distributors, third party logistics providers, transporters, etc). The winners are quickly identifying the gaps in existing performance to determine where and how to leverage on new structures, new technologies and new capabilities to deliver the level of performance necessary to serve the intensely competitive marketplace.

With a clear understanding of performance priorities and existing gaps, along with the corresponding value proposition, companies can define their respective journey plan which would be essentially based upon a common platform and a company wide approach. Successful implementation of such a journey plan would lead a company to become a truly integrated enterprise which is the basic foundation for supply chain excellence.

Examples do exist where the entire system gets reoriented depending upon the magnitudes and qualities of the orders placed on the industries.

Practice of agile manufacturing has therefore become, of late, an absolute necessity in view of the increasing demands of customer-oriented products. Hence, the entire system needs to be made agile and flexible to meet the challenges faced by the Indian industrial house.

In view of the welfare of the workforce which has been a vital component of the business, a system for long term education and training of the workforce is essential. Whereas the middle level and top level workforce can be deputed to institutions for higher learning as a part of quality improvement programme, the labour class will have to be provided in-house education and training programmes. Organizational pattern must be flexible enough to provide perfect cooperation among the three components of agile manufacturing: management, technology and workforce.

(i) **System parameters under the technology component** that need to be integrated with management and workforce components of the system:

To achieve balanced integration of management, technology and workforce, an interdisciplinary design methodology must be adopted to bring the concept of total enterprise design. Emphasis must be laid on strategies to enhance skill and knowledge of workforce and not on replacing old technology by new technology. Technology deployment is as important as technology development. Technology deployment must be viewed as a criterion for technology development. Research into technology must involve all the three components to arrive at an appropriate technology most suitable to the system.

This appropriate technology could be either arrived at by human resources within the system or borrowed from an outside agency. Energy conservation and waste minimization and management are the two parameters which occupy top priority especially in view of energy crisis and threat to environment and resources.

(ii) **System parameters under the workforce component** that need to be integrated with management and technology:

The ultimate aim of the entire business system being satisfaction to the customer, it is essential that the workforce should be built up. Quoting the words of **Mahatma Gandhi:**

- A customer is the most important visitor on our premises.
- He is not dependent on us, we are dependent on him.
- He is not an interruption on our work, he is the purpose of it.
- He is not an outsider to our business, he is a part of it.
- We are not doing him a favour by serving him, he is doing us a favour by giving us an opportunity to do so.

Agile manufacturing should provide the desired flexibility to manufacturing houses, in view of our (Indian) liberalized economy and the stiff global competition posed by multinational companies and units started by them since products manufactured by Indian manufacturers should meet international standards in respect of quality and price.

As we see, Indian manufacturers have been facing a century long decline in industrial competitiveness while other nations like Japan, South Korea and the thirteen member states of the European community are in the process of uniting their national markets to establish prominent competitive positions for their own industries. In this background, if India lingers, its manufacturing capability will continue to decline, and it will be blocked out of many world markets.

Gifted with so many precious advantages to progress indigenous industry, pollution prevention and elimination and upkeep of ecological balance including environmental protection, agile manufacture in India is one of the invaluable inputs as it will fully address the importance of the country's competitive position, namely, making a dent as a world leader in engineering, information, technology, etc., and widening its diverse supplier bases that constitute a major resource for cooperative ventures in an agile and concurrent manufacturing environment. This will not only create a significant advantage but help implementation of agile manufacturing (partially/fully) in terms of:

- **Enhanced Management Aspects:**

Adoption of

- Vision Based Management
- Open System Architecture
- Customer Responsiveness
- Dynamic Diversification Capabilities
- TQM
- Enterprise Integration
- Enabling to Build Employees as Vital Asset
- Preventing Wastage in Management

- **Enhanced Technology Aspects:**

Build Leadership by:

- Right First Time in Design and Manufacture
- Concurrency in all Activities
- Green Technology Usage
- Environment-friendly Products Manufacture
- Short Cycle Time

- **Enhanced Workforce (highest to lowest) Aspects:**

Helping to Boost up:

- Skilful, Knowledgeable and Learnable Workforce
- Technology Awareness and Leadership
- Highly Motivated and Wilful Workforce experiencing Life-Long Learning
- Empowerment of Individuals Working in Teams
- Accessible and Useable Information Utilities

(iii) **System parameters under the management component** of agile manufacturing which need to be integrated with technology and workforce are as follows:

Traditional management accounting to provide flexibility which is vital for agile manufacturing – activity based accounting provides this desired flexibility. Since the technology that is adopted by the system is to be user-oriented, it is essential that the user must have control over the technologies that will be adopted by the management. This can be brought about effectively by organizing interactions between the user and the workforce in an industry.

In conclusion, the following are the inherent invaluable advantages of agile manufacturing:

(a) Agility in manufacturing
(b) Production of cleaner green products/components
(c) Customized products

The resources such as materials, energy, water, labour, machines, etc. required to manufacture components and products are conserved due to limitation to the required specifications and orders of customers leading to perfect sustainability because nothing will be manufactured in advance and kept for sale.

8

Agile Manufacturing: The Advent of a New Frontier

As outlined earlier, to succeed in agile manufacturing, it is necessary to use people to transform their knowledge into ideas for new and improved products and processes and discover better ways of doing things. This change of emphasis requires the development of a whole new approach to manufacturing.

Agile manufacturing is often misunderstood as a quick fix or a prescription. In fact, it is none of these. Agile manufacturing is not flexible manufacturing nor lean manufacturing nor computer integrated manufacturing. It is a business opportunity which requires creativity and innovation to turn it into enhanced competitiveness and profitability. The business drivers of agile manufacturing are primarily concerned with achieving quantum leaps in competitiveness and performance and making life difficult for competitors.

It is important and necessary to develop a good understanding of what agile manufacturing is about before embarking on the development of its capabilities.

The key to agility depends upon the fundamental concept of integrating management, workforce and technology into a coordinated whole. Agile manufacturing is primarily a business concept. This is what delivers agility and competitive advantage. Agility is not a function of one single dimension. Agile manufacturing has several generic features which have to be designed into the manufacturing infrastructure.

8.1 FOUR CORE CONCEPTS AND ASPECTS

The four are mutually interdependent. The four core concepts of agile manufacturing are:

- Need to develop a strategy to become agile
- Strategy to exploit our agility to achieve competitive advantage
- Strategy to develop agile manufacturing should focus on ways to achieve integration of management, workforce & technology
- To achieve this integration, need to develop strategy for interdisciplinary design methodologies.

In order to remain competitive, continuous improvement in all areas is necessary to keep an edge over competitors. Change, therefore, must be continuous.

Given a process of continuous change, the question is, how best it should be to initiate and to manage it? This leads to the idea of transitional systems that can be regarded as opportunities for tuning in (getting in touch with new situations to get a feel of their texture), working through (understanding the implications and facing up the pain and emotion of upheaval) and designing (creating the new situation).

The concept of transitional systems can be of help to develop a new manufacturing paradigm. Embedded in the idea of transitional systems is the concept that one should clearly understand the existing culture and the implications of the proposed changes before embarking on the process of change and begin to design new systems.

Agile manufacturing implies revolutionary changes to enterprises. One should not be satisfied making small adjustments, but reinvent the manufacturing enterprises. There is need to understand the reasons for past failures and make sure that these are not repeated as mistakes. A crucial factor underlying past failures is outdated tradition since the methods and approaches used in the past have now largely outlived their usefulness and should be abandoned. We should understand the wide ranging nature of the changes that lie ahead and also that we all have to change. Change is not just something that others have to go through. What we need are new methods, processes and values; move beyond quick fixes, panaceas and prescriptions and ask fundamental questions about what to do and why, how to do it, and when. Thus, agile manufacturing implies a paradigm shift. The key to success lies in understanding the change processes and in professional and innovative management of that change.

Understanding the limitations and applicability of the traditional management accounting techniques is crucial to success in agile manufacturing. This is so because management accounting systems influence behaviour of people throughout the enterprise; help to preserve traditional approaches to manufacturing, values and beliefs; and are a key leverage point for long-term changes. However, changing the management accounting system brings its own problems. The traditional investment appraisal methods, on their own, are not well suited to supporting strategic investments in agile manufacturing. To quantity investments in agile manufacturing, we should not give up quantity benefits. This implies that traditional approaches need to be enhanced. Use of traditional absorption costing for management decision-making which involves using product cost information is often inappropriate, because of the cross-subsidizations introduced by this method. Thus, traditional performance measurements and control systems, which are largely based on financial information, are not appropriate for agile manufacturing.

Technology leaves little room for the development of personality, for achieving personal aspirations or for exercising the thinking premature abilities. This represents a waste of human resources, which are some of our most precious assets.

In industrial plants there are many activities where the skill and knowledge of people are required before full benefit can be achieved from the technologies used. The machining of metals is a good example. It is possible to construct a mathematical model of the machining process which can be used, given the availability of certain data, to calculate the optimum way of cutting metal.

This, however, is only a theoretical possibility. In practice, such a mathematical model is likely to cope with only the simplest and most carefully defined situations. The reality is that the data used in such models are subject to large degrees of uncertainty. The cutting process itself is not fully understood and cannot therefore be accurately modelled and the work is subject to a large number of unpredictable disturbances.

This is a very simple example of the type of situation where the benefits of people's skills and experience can be used most effectively. We can combine people and computers to produce a result which is greater than the sum of the component parts. It is in such situations that we can use to good effect the abilities of people so carefully developed by educators.

Dependence on people's skills and abilities is necessary but often only in a tacit way. In practice very often, satisfactory facilities are not developed to enable people to achieve a human-machine symbiosis. Typically, the need for human skill is seen as a defeat and the course often taken is to 'eliminate' the people. In reality, however, people are asked to do the tasks which are not automated and to provide support when the systems fail to function as desired.

This process of job destruction is counterproductive. It leads to alienation, lack of motivation and commitment and an unwillingness to cooperate, change and improve. Thus, traditional design practices are completely at variance with what we are trying to achieve in agile manufacturing.

In today's competitive environment, where focus will be on agile manufacturing, traditional arguments are no longer relevant. Hence, there are good economic reasons for designing and developing technologies that collaborate with people's skills, rather than technologies that reject them.

In agile manufacturing, one should not continue to waste human resources and with practices that have contributed to the creation of many people related problems. We need to develop a more friendly technology that collaborates with people's skills. The question however remains: how do we set about designing this type of technology?

Traditional concepts of organization and control, and traditional technological and design paradigms were shaped by a different set of circumstances to those in an agile manufacturing enterprise. Here the way we do things must be fundamentally changed if we are to succeed with agile manufacturing.

Traditional manufacturing plants are built around organizational control, technological and design paradigms that have ceased to be relevant.

The **Taylor Model** is deep rooted in our management and engineering culture. Before we can change this culture, we need to fully understand the elements of the Taylor Model. This model is not suited to modern manufacturing and competitive environments. It represents a barrier to achieving agility and it is in direct opposition to ideas such as continuous improvement.

We must resolve the long-standing problems of alienation, lack of motivation, and so on within enterprises. The model is the bottleneck that comes in the way of a successful resolution of these problems.

The role of technology is shifting from work automation to supporting complex decision-making. This requires new concepts. We need to develop a new technology design paradigm.

It is not possible to develop a truly agile manufacturing enterprise using an outdated approach. What we need is a design method that does not reduce the complexity of the problem through systems decomposition, but which allows us to work with the complexity. We need to fully understand and explore the complexity that arises from the interrelationships that exist between subsystems and to deal with the systems as a whole, and not as isolated subsystems and components. This needs application of the concept of total enterprise design, which is based on the foundations for world-class manufacturing systems. These foundations should be fully understood and seen as interacting subsystems. The key to total enterprise design is keeping the full complexity of the interaction in view while dealing with specific subsystems.

Hence, the approach must be a holistic one, away from quick fixes and panaceas. We need to understand the key elements of the systems theory, such as emergent properties and equifinality, before embarking on the application of the systems theory to agile manufacturing enterprise design.

Enterprise design process is therefore a key component of agility. In adopting the spiral approach, we not only seek to identify and eliminate risks, but to achieve the intertwining of problem solving and problem seeking and the rapid adaptation of strategies and systems to the continuously changing needs of the marketplace.

But it is also possible to achieve more than this. The spiral approach provides an opportunity of reintegrating designers and users, bringing the users' perspective, back into the process. In short, we can start the process of change, long before that process becomes evident in the form of new technologies, new machines, new organizational structures and new working practices.

Stage-wise models of design are hierarchical and rapid. They lead to an inherent lack of agility. To become agile, these outdated design strategies must be eliminated while adopting more appropriate ones, without losing structure and discipline. The spiral approach, in principle, has the necessary structure and discipline built into it. What is primarily lacking are the tools to support an enterprise design process based on the spiral strategy.

Traditional design theory separates **problem solving from problem finding**. In more modern theories of design, these two aspects are seen as intertwined. Many models of design process are based on stage-wise methods which do not correspond to the reality of complex design situations.

In some situations, different types of design process are needed, such as incremental strategies, adaptive strategies or spiral strategies. These approaches are likely to be more relevant to agile manufacturing than stage-wise strategies.

The psychological perspective of a responsive agile manufacturing enterprise is represented by the 'people'. This broadly covers two areas. The first is health and safety and interface and work space ergonomics and the second is the skills, knowledge, needs and motivation of individuals. The organizational perspective can be represented by an 'organization' box. This again broadly covers two areas. The first is the structure and role of teams and the second is the network of teams covering relationships and interaction between teams, which includes those from outside the plant or the company, as in the case of supply chain considerations.

The direction taken from 'organization' and 'people' to 'technology' is important. This is as a two-way relationship between organization and people on the one hand and technology on the other. This relationship shows that organization and people considerations can be used to shape technology, and vice versa. In practical design situations, there will be trade-off between organizational, people and technical considerations.

The fundamental requirement for achieving agile manufacturing is the ability to design manufacturing enterprises, systems and technologies in an interdisciplinary manner. The benefits that can be obtained from such an approach include improved systems design leading to faster implementation and avoidance of potential problems. Of equal importance, however, are the opportunities that open up for developing new ideas and new solutions by exploring the areas that lie between disciplines. However, achieving interdisciplinary design is not easy and will take a long time to become normal practice. Success and failure in systems design activities has to be judged against external benchmarks, as well as by internal criteria.

Systems design processes are a competitive weapon. If we can improve them, and develop capabilities which competitors do not possess, or will find hard to imitate, then success proves to be harder.

The complexity of agile manufacturing design problems necessitates an interdisciplinary approach to design. This interdisciplinary design should not be confused with multidisciplinary design. The two terms are not synonymous. The aims should be not only to apply insights and knowledge from several domains, but also to explore the areas that lie between disciplines. It is in these areas that we may find new ideas and new and original ways to step forward.

Design is affected by the values that people hold. It is not a purely rational process. Therefore, design method alone cannot lead to improvements in design process. We should understand the differences between the surface characteristics of computer-based technologies and their deeper characteristics. Technologies can constrain organizational choice. When we design technologies, we are designing organizations and jobs.

As explained earlier, it is important to note that an approach to management accounting and investment appraisal suitable for agile manufacturing environments should be broad, and the investment appraisal should be seen as an analysis tool to support design of the agile manufacturing enterprises. The analysis technique used for appraisal should be deployed concurrently with design, to help shape designs and design choices. The concept of activity-based costing is a potential tool to support decision-making. Regrettably, its potential is not fully exploited. We need to develop new management accounting and investment appraisal techniques for agile manufacturing, using new concepts such as ABC. We need to improve our capabilities at quantifying so called intangibles, but also to deal with softer issues such as motivation and skill base.

Focusing on cost reduction alone is not appropriate. We need also to look at revenue generation enhancement through the development of agile manufacturing capabilities.

Analytic and strategic appraisal methods should be used to complement financial appraisal methods in a spiral approach. The aim is to use the techniques to support design. Investment appraisal should not be viewed as a hurdle to clear, once the design options have been fully developed.

Knowledge-based systems containing knowledge from a number of knowledge domains could thus be used to assist people with their work. These systems, however, would not only provide a transactional operating mode, if this mode was considered appropriate, but would also be capable of operating in supervisory, commentary and learning modes. Such systems would be advising people about the consequences of their decisions and adding to their knowledge, enhancing their skills. Skills and knowledge enhancing technologies represent a new paradigm for the application of computers in manufacturing. These technologies are a key component in agile manufacturing.

Observations

These paragraphs describe an example of skill and knowledge enhancing technologies. So far, we have only considered skills and knowledge that to a large extent already exist and the application to process-oriented functions. We should also consider the development of skill and knowledge enhancing technologies for skills and knowledge which are at the moment not perceived as part of our people's work. For example, we could think about several other areas where these technologies might be applied, such as:

- Innovation skills–Skills related to using creativity to find better ways of doing things
- Team skills–Skills required for effective cooperation and collaboration, both within the machine tool system and with other subsystems.
- Communication skills–Skills required to effectively transfer insights to other people.

A key step in the design of skill and knowledge enhancing technologies is to undertake a skills, knowledge and task analysis. The focus of the design should be on making use of people's skills, but in a way that is satisfying to the users, and without over stressing them. User control over the technologies is an important issue, as in the capability to accommodate individual working styles, and so on. Research in the area of agile manufacturing needs to be addressed in a more interdisciplinary way.

General lack of understanding, added with bias towards non-disciplinary research projects, has made it difficult to undertake interdisciplinary projects. There are barriers imposed by tradition, that frustrate attempts to develop agile manufacturing. It seems inevitable that research must ultimately become more interdisciplinary.

We are facing a paradigm shift away from the **Taylor Model** to a new paradigm which we call agile manufacturing, in which people's attributes, such as their skills, knowledge, innovative ability, and so on play a central and leading role. This change of paradigm demands a new way of operating, new values and a new type of computer-based technology which is designed to make people's skills and knowledge more productive and effective. However, the deep-rooted nature of the old paradigms make it extremely difficult to recognize the importance and commercial potential of agile manufacturing and to develop the skills and knowledge enhancing technologies required to make it a success.

There is also an urgent need to address research and development in this interdisciplinary area, for it is likely that important new discoveries and interesting insights, leading to further innovation, will be generated by working in the areas between disciplines. Someone has to pioneer the development and commercial exploitation of agile manufacturing. Overcoming the legacy of the **Taylor Model** is a major barrier to progress.

There is danger that agile manufacturing may be interpreted as a technological concept and will lead to the generation of more and more technology, but not to the development of the capabilities to deploy the technologies. The development of agile manufacturing requires an integrated approach as said above. Piecemeal and fragmented research by different disciplines will only hinder the development of the concept.

8.1.1 The Legacy of the Taylor Model

The big issue is not whether we should get involved in the development of agile manufacturing, but to what extent our own existing beliefs, goals, objectives and methods will continue to be shaped by the old manufacturing paradigms. We can easily adopt the label of agile manufacturing, but can we easily discard the thinking behind the **Taylor Model** which represents an engineering and management culture that has developed over the past 200 years? Will our work be technology driven and non-disciplinary or will it be more balanced and truly interdisciplinary?

We should focus primarily on some recommendations for future research in the area of agile manufacturing, based on an interdisciplinary approach and on a balance between organization, people and technology.

8.1.2 Need for Strategic Integrated Approach

Old approaches to manufacturing, which are largely based on separate and isolated functional departments, have outgrown their usefulness and have become a competitive liability. This fragmented approach to manufacturing is slowly being replaced by an integrated approach, the keywords being empowerment, simultaneous activities, coordination, cooperation, sharing and team working. However, this message has still to penetrate into the operating philosophy of governmental departments and other governmental type organizations that fund research activities.

No organization can expect to deal adequately with the development and implementation of measures to support the development of agile manufacturing, without first adopting a new approach itself. Many of the issues involved need to be approached in an inter-disciplinary manner. However, human factors research programme on its own or within a technical research programme, is not adequate. What is also required is to integrate these aspects into the technology development research.

8.1.3 Awareness of Scope and Importance of Agile Manufacturing

People still do not fully appreciate the scope and importance of organization and people issues in agile manufacturing. They have a tendency to think of human factors, which we perceive to be primarily an organizational question or a quality of working life issue or associate with training or ergonomics of human-computer interfaces. Moreover, while there is increased awareness of the need to improve the management of change, people tend to view this solely as a question of how best to deploy more technology.

The potential for improvements in profitability and competitiveness that can be derived from tapping into and mobilizing people's competencies, i.e., their skills, knowledge, judgement, creativity and experience is still not properly understood. The need to remove organizational layers is widely talked about, but it is not clear whether the motivation for doing this is related to cost savings or in achieving broader objectives.

For example, there is a belief that information technology has made a large number of our middle managers redundant, because the monitoring and controlling actions that they perform can now, to a large extent, be done with the aid of computers. The reasons why we have so many middle managers in the first place is because of our command and control mentality that prevails in our approach to management. When we start to empower people, many of our middle managers are no longer required to fulfil control and monitoring duties. Also, the role of computers has to change from one of monitoring and control, to one of information and communications support. Thus, the purpose in removing organizational layers is not just to flatten the organizational hierarchy. We should see such moves as a first step in the complete dismantling of the command and control organizational hierarchy.

8.1.4 Education and Training of Engineers for Success

The education and training of managers and manufacturing systems engineers is an area that needs attention. We need to consider the skills and knowledge that our managers and engineers should be acquiring from undergraduate courses. At the moment it seems that the education of most managers and engineers is firmly rooted in the production engineering culture of the 19th century.

Education and training will be a key area for the success of agile manufacturing and our exploitation of research results.

First, agile manufacturing is not just a question of addressing issues such as organizational aspects and psychological topics in isolation. These have an impact on technology development, which should now be addressed in an interdisciplinary way. Second, the success of agile manufacturing depends to a large extent on the availability of well educated and trained, highly skilled people throughout the enterprise.

New knowledge and new methods of working are needed. Most of our managers and engineers do not, at the moment, have enough knowledge or sufficient skills to work effectively on interdisciplinary projects.

Demographic changes also imply that great emphasis should be placed on the retraining of older people. The increasing emphasis that younger people place on quality of working life will mean that continuing training will have to become the accepted norm if we are to attract and retain younger people.

Universities and colleges will have to develop appropriate education and training for managers, engineers and technicians. Emphasis should be placed on:

- Education and training for lifetime learning
- Developing social, teaming and communications skills for group working
- Learning by doing
- Increasing links between industry and educational establishments
- Open learning approaches
- A psychology of learning driven approach to the development of multimedia learning technologies
- The training needs of small to medium size companies
- Broadening of skills and knowledge

There is need to develop interdisciplinary manufacturing systems engineers; and to develop design team structures to support interdisciplinary design. University undergraduate courses will have to be revised and training provided for professors as well as engineers working in industries.

Management competence also needs to be improved. Managers must be helped to develop a stronger people orientation, and they need to become more familiar with modern management science theory and practice. Training is also needed to help managers to deal with change management and develop styles of management in keeping with the new emphasis on trust, empowerment and team working.

The training of people for the computer age also requires more serious attention. Empowerment implies increasing competence. We do not need hierarchical managers, narrowly oriented engineers, semi-skilled machine operators and single skilled people. We need multi-skilled, computer competent people. Everybody must acquire team working skills and become more cooperative.

While the technical specialists will still be needed, many manufacturing systems engineers should try to become manufacturing strategists and systems integrators, with a broad understanding and knowledge base.

All these changes are problematic. The whole of our educational system is rigidly divided into disciplines. We do not have many interdisciplinary departments in universities. When we visit an industrial engineering department, we tend to find only industrial engineers, or other related engineering disciplines. There are a few, if any, non-technical members of staff. When non-technical issues are taught, this is usually by people from other departments. The issues that lie between the various disciplines are almost never explored. This, unfortunately, reinforces our culture of non-disciplinary methods and thinking. If we want to change the way that managers and engineers think and act, then we have to change their education. But before we can do that, we have to change the educational system that reflects the very structures and divisions of knowledge that agile manufacturing has made redundant. And we must also re-educate our educators.

8.2 AGILE MANUFACTURING RESEARCH

8.2.1 Strategic Development of Knowledge, Competence of Skills and Judgement Issues

It is clear that agile manufacturing requires our research community to make a strategic shift away from the development of automation technology, especially in the area of decision-making, towards using technology to support people. The emphasis needs to be on developing skill and knowledge enhancing technologies, rather than skill substituting technologies. We should develop technology which leaves room for the exercise of people's skills, knowledge and judgement. Technology should be used to make these distinctive attributes of people more productive and effective. Our efforts to reproduce these unique characteristics in computers are likely to become more and more difficult as the complexity of the decision-making environment continues to increase. Most of the easy problems in decision-making have now been dealt with. The ones that are left are the difficult ones, which, if they are to be tackled successfully, will require a completely different approach.

It is also clear that we are unlikely to be able to develop skill and knowledge enhancing technologies on our own. Part of the expertise wanted lies outside our area of knowledge. We should therefore collaborate with organizational scientists and psychologists and address the development of technology as an interdisciplinary subject.

As pointed out under the heading 'Observations', here also, there is a growing need to adopt a broader approach to technology development and deployment. This should not be a purely technical one. Development projects should be expected to employ a holistic approach, taking into account all the factors necessary to bring about change. These include skill and knowledge enhancement issues, technological requirements, management and organization, training and motivation of staff and technology deployment. Projects should be balanced, addressing several elements including organization, people and technology.

It is important that we change our assumptions about organization and people issues. These are not secondary matters nor are they issues that should only be dealt with at the implementation stage. In agile manufacturing, they permeate all aspects of manufacturing enterprise design, including research and development. We should assume, therefore, that all our projects have an organization and people dimension. We should make it an acceptance condition that all research proposals identify the organization and people issues and specify how these will be dealt with, or alternatively, demonstrate that organization and people issues are not relevant. Our project evaluation teams should include experts with knowledge and experience in this area. We need to develop criteria and guidelines to help research and development people to deal with these aspects. These criteria and guidelines should address issues such as organizational structure, the role of people, skill and knowledge enhancement, human-computer interaction, ergonomics, motivation, job satisfaction, training, health and safety.

Those in manufacturing who become involved in research and development proposals seeking government support should be expected to give details of their organizational structures and practices. Organizational structures and procedures should be justified and organizational redesign objectives should be started. This information should be taken into account during the evaluation of any project proposal.

Once a government funded project is approved and running, project monitoring and formal reviews should pay particular attention to the way that organization and people issues are being addressed by projects. Probing questions should be asked. The importance that is attached to interdisciplinary working should be made clear through the performance criteria used to evaluate projects.

8.2.2 Specific Research Issues to Support Enterprise Design, Manufacturing Research and Technologies

Research needs to be undertaken on a wide range of topics. Consideration should be given to the following:

- The concept of agile manufacturing enterprise design. There is a need for methods and tools to support an interdisciplinary approach, with special emphasis on combining top-down and bottom-up methods, holistic approaches, organizational simulation, appropriate and selective use of technology, increased user involvement and rapid prototyping. These are all relevant which will be required to support a spiral approach to the design of agile manufacturing enterprises.
- More emphasis on developing decision support systems that expand the range of possible decision alternatives. Avoidance of a single minded emphasis on the speed with which a decision is made with more emphasis on the quality of the decisions.
- The need to address research in manufacturing as a whole, not just research into new manufacturing technologies. The need to develop new management practices, technology deployment techniques, new reward systems, more sophisticated financial justification techniques, new management accounting methods, and so on to support agile manufacturing.
- The development of skill and knowledge enhancing technologies focusing on applying insights from the organizational and psychological sciences. Condition monitoring, database management and networking, applications of genetic algorithms, neural networks and so on to support decision-making, vision systems, workshop-oriented programming of robots and machine tools and computer supported cooperative working.

- Technologies that will support a shift in market conditions towards one-of-a-kind production (OKP). Agility in manufacturing is a key aspect.
- Technologies to support group working. Increasing emphasis on team or group working. Computer supported cooperative work.
- Technologies to support the learning company or organization.

8.2.3 Future Developments of Knowledge: Enhancing Technologies

The preceding paragraphs have outlined several developments which constitute skill and knowledge enhancing technologies. There are two particular areas where further development can be expected. One is the use of artificial intelligence (AI) technologies and the other is in the area of cooperative working.

AI is a rather unfortunate term for it conjures up emotive issues of intelligent machines taking over intellectual work that has hitherto been immune from automation. It also generates philosophical discussions which centre around the possibility of developing computers that think, which leads to endless speculations largely based on opinions. These roughly divide into those who believe that computers can be made to think and those who do not. Both sides in the argument have a narrow focus on the world. It is not our aim to promote these new technologies as panaceas, but neither do we wish to adopt an anti-technology or anti-AI position.

We should prefer to adopt a view of AI based on the concept of programming paradigms, because it takes account of established programming practices known as the procedural paradigm which many people will be most familiar with. It is based on defined routines which are executed in some predefined ways depending on the inputs. We have seen over recent years the emergence of several advanced programming paradigms, such as, in modern programming environment, we should not restrict ourselves to the exclusive use of any particular paradigm because there is no single best programming paradigm for all purposes. In the case of complex problems it is, in fact, necessary to combine several paradigms to fit the various tasks. For example, what can be difficult to achieve using a rule-based paradigm, might well be quite easily achieved using the procedural paradigm and vice versa. We should view these paradigms not as a threat to the skill oriented design paradigm, but as tools to support the achievement of the sought after goals and functionality. For example, we could embed a rule-based component in some software in such a way that the users or an outside observer might never know this, unless told.

Within the framework of open systems, advanced programming paradigms can be used to support people rather than replace them, and this requires that we establish the extra skills and knowledge people require in order to work more effectively. We must also select an operational mode that leaves people with the freedom to act in the way that they want.

8.2.4 Future Generation Skills and Knowledge Enhancing Technologies

Traditionally, design of technology has focused on process-oriented tasks, usually attempting to automate people out of the decision-making process. People's roles have normally been restricted to handling tasks that cannot be automated and dealing with unpredictable and unexpected events. Integration has usually been restricted to interfacing between computer based technologies and sharing data within computer applications.

The design of skills and knowledge enhancing technologies is based on a wider understanding of integration, namely:

- People integration (people communicating and cooperating with people).
- Human-computer integration (people integrating with computers).
- Technological integration (machine interfaced with machine).

People have become all too familiar with the technological integration perspective, which might more appropriately be described as the technical interfacing perspective. One aspect is the human networking organization, which involves many linkages between people, groups of people, functions, and so on. In each enterprise there will be a unique set of critical linkages, which will change with time as the nature of the business and the market changes.

The way in which people's integration aspects of these critical linkages could be developed and in what way could human-computer integration and technological integration could be developed to support the people integration aspects of the critical linkages depends upon defining the attributes (skills, knowledge, and so on) that would be needed and then on how the decision to be made more productive and effective through the use of skill and knowledge enhancing technologies. Immediately, the nature of future generation skills and knowledge enhancing technologies becomes clear.

Presently, most computer-based technologies are process task oriented, that is, they are primarily concerned with the tasks associated with material transformation such as programming and operating a machine tool and this is where the focus of attention has been placed on the development of skills and knowledge enhancing technologies. Future generations of computer-based technologies will also be process task oriented, but there will be additional technologies, some of which go beyond the traditional process task oriented technologies. These will include technologies to support experimentation and learning activities and technologies to support the linkages within the human networking organization.

8.3 EXPERIMENTATION AND LEARNING TECHNOLOGIES

8.3.1 Selection of AI (Artificial Intelligence) Technologies

These technologies will be concerned with:

- Supporting the process of continuing improvement
- Helping users to extend their understanding of problems
- Providing advice in adjacent areas of expertise
- Supporting experimentation and learning activities
- Informing users about consequences of proposals and decisions
- Helping to identify potential problems and failures

These technologies should not be confused with computer-aided learning and training technologies. Here we are talking about technologies that will help people to extend their understanding of a problem domain through the use of conceptual and procedural guidance or support experimentation and learning exercises that are particularly focused on finding better ways of doing things.

8.3.2 Linkage Technologies

These technologies will be concerned with supporting the process of forming and developing critical linkages within the enterprise. They will:

- Support the identification and formation of critical linkages
- Help empowered people to contribute to team activities
- Support inter-group and intra-group activities
- Identify core competencies
- Support inter-enterprise and intra-enterprise innovation networks

The emphasis here is on using technologies to support empowered individuals in their effort to join, form or participate in teams and to make people's contributions to the work of these teams more productive and effective. This is what is required to succeed in agile manufacturing.

9

Agile Manufacturing: A 21st Century Paradigm

The term agile manufacturing came into common usage with the publication of the **Report 21st Century Manufacturing Enterprises Strategy (Lacocca Institute, 1991).** Agile manufacturing has been used to refer to other concepts such as flexible manufacturing and lean production, as though synonymous.

9.1 DEFINITION OF AGILE MANUFACTURING

Before going to the definition of agile manufacturing it is necessary to understand some of the terms like agility, lean manufacturing, etc; and concepts linked to agile manufacturing.

Agility is defined in dictionaries as quick moving nimble and active. This is not the same as flexibility, which implies in the manufacturing sense, adaptability and versatility. It is now an accepted assumption that flexibility is a requirement for the competitive markets or today, but on its own, will not deliver agility. Flexibility should be regarded only as necessary condition for agility.

Lean manufacturing (Womack, Jones and Roos, 1990) is so called because it is concerned with doing everything with less (Jones, 1992). In other words, the excess of wasteful activities, unnecessary inventory, long lead times, and so on have been cut away through the application, of: for example, Just-in-Time Manufacturing, Concurrent Engineering, Overhead Cost Reduction, Improved Supplier and Customer Relationships and Total Quality Management. However, Lean manufacturing is not the same as Agile manufacturing, because Leanness and Agility are two different things.

Our new vision of manufacturing will be based on a system perspective technology, organization and people, tied to clear business vision and goals. This will help to understand the full complexity of designing a 21st century manufacturing enterprise and the way that the past mass production paradigm still limits our thinking today. Most of all, this system's perspective will help us to see how to approach the task of designing an agile manufacturing enterprise. The issues addressed are: defining what we are about, understanding the present and how it limits our progress and the means by which we will bring about agile manufacturing.

In future, the traditional approach will be replaced by a more balanced and interdisciplinary and holistic systems approach of organization, people and technology, the creation of a learning environment and a focus on cost, quality and flexibility.

Agile manufacturing can be considered a structure within which every company can develop its own business strategies and products. The structure is supported by three primary resources. Underpinning these three primary resources is a methodology of integration.

- Innovative management structures and organization,
- A skill base of knowledgeable and empowered people
- Flexible and intelligent technologies.

The competitive environment of the future, as this method is usually appropriate in circumstances where there is a high degree of stability (Minitzberg, 1990). In circumstances where market conditions are highly dynamic and risk is high, this approach needs to be replaced by a more evolutionary method of strategic management.

9.2 KEY POINTS

- Agile manufacturing implies revolutionary changes to our enterprises. We should not make small adjustments. We should reinvent our manufacturing enterprises.
- We should understand the reasons for past failures and make sure that we do not repeat these mistakes and these should be documented and should be available for everyone in the company for reference.
- A crucial factor underlying past failures has been an outdated tradition. The methods and approaches used in the past have now largely outlived their usefulness and should be abandoned.

The Iacocca Institute report also describes the concept of agile manufacturing in terms of several company scenarios. The teams of industrialists responsible for producing the report produced a set of competitive foundations.

A conceptual framework for agile manufacturing has already been presented. (Ref. Section 8.1)

9.3 CONCEPTUAL ASPECTS OF AGILE MANUFACTURING

To retain and increase our competitive position in world markets and to respond to windows of opportunity, we should be able to take full advantage of our potential for increased productivity, better quality and greater flexibility provided by more responsive and new forms of organizational structures. We also need to make better use of human skills, knowledge and experience, and exploit the power of modern computer-based technologies. To achieve all these goals we need skilled, cooperative and motivated people. Our traditional management hierarchies and methods of control must be eliminated. More participation is needed by people throughout the enterprise, in planning, designing and implementing new technologies and systems. Technical systems also need to be designed, not just to meet economic and technical goals, but to satisfy organization and human requirements.

Agile manufacturing is a modern manufacturing concept that embodies these ideas. The importance of the human role in agile manufacturing is stressed. However, agile manufacturing should not be seen as a collection of management and organization techniques, in which technology has no major role. A wide range of human abilities and characteristics are required in an agile manufacturing environment, and we must focus more on people issues, both to make full use of all our peoples abilities and to create a good working environment for everyone.

To bring agile manufacturing to fruition (completion, end result), we have to overcome our belief that technology alone can provide the answers to all our problems. We should treat non-technical problems as technical problems. A much more sensible approach is needed if our manufacturing companies are to compete in world markets. Reduced costs, improved quality, greater flexibility and rapid response capability are needed, without trade-off. We cannot do this with technology alone. We need a balanced manufacturing response, not a technological one. This demands the development of a new and broader vision of the factory of the future: a vision with a future. Our new vision should be technology-oriented manufacturing which leads to a competitive role and with an aim to create an environment for the exercise of human skills, judgement, creativity, knowledge and ingenuity and to make full use of modern computer-based technologies. These two goals are essential requirements for future success in manufacturing.

9.4 AGILITY DRIVE: A STRATEGIC APPROACH

We can only benefit from agile manufacturing if we have a strategy of agility which will allow us to formulate a change, the plan to implement agile manufacturing and be competitive.

9.5 COMMENTS

The idea now is that, after spending the past 200 years trying to get people out of manufacturing systems, it will now be necessary to start putting them back. We need to use people to transform their knowledge into ideas for new and improved products and processes, and discover better ways of doing things. This change of emphasis requires the development of a whole new approach to manufacturing. Consideration should be given to the important concepts needed to bring about an agile manufacturing system.

- It is important not to confuse agile manufacturing with flexible manufacturing, lean manufacturing or computer integrated manufacturing.
- Agile manufacturing should primarily be seen as a business concept.
- Agile manufacturing could be seen as another buzzword (catch phrase), a quick fix or a prescription. In fact, it is none of these. It should be seen as a business opportunity which requires creativity and innovation to turn it into enhanced competitiveness and profitability.
- The business drivers for agile manufacturing are primarily concerned with achieving quantum leaps in competitiveness and performance, and making life difficult for our competitors.
- It is important and necessary to develop a good understanding of what agile manufacturing is about before embarking on the development of its capabilities.
- Agile manufacturing has several generic features which have to be designed into the agile manufacturing infrastructure.
- Agile manufacturing requires the integration of organization, people and technology into

a coordinated whole. This is what delivers agility. Agility is not a function of one single dimension.

- Multi-skilled and flexible people
- Team working
- Empowering of all the people in the enterprise
- Knowledge management
- Skill and knowledge enhancing technologies
- Continuing improvement involving all people
- Change and risk management

It should be stressed that these generic features are not a prescriptive framework for agile manufacturing, nor are they intended to be a prescriptive solution to the various problems that beset manufacturing enterprises. The needs of manufacturing companies are not generally amenable to prescriptive solutions.

One of the problems with the emerging competitive environment is that there are so many different types of manufacturing business; high volume and low variety, low volume and high variety, process, batch and jobbing and long and short life cycles and most businesses trying, within this complex environment, to differentiate and compete in a wide range of markets. This means that each business has its own specific set of critical competencies. As a result, there is a wide variety of them, and no prescriptive approach is likely to provide adequate solutions for a wide range of businesses.

This implies that the generic features outlined above are to be highly tailored to specific needs of manufacturing companies or even product families and should be considered only as a model that provides us with a guide to the sort of building blocks needed to develop an agile manufacturing enterprise.

- Ease the transition from old to new work methods.
- Create a flexible system which will allow our skilled users to develop working methods that are appropriate to each situation.
- Leave room for users to develop and practice their skills and to contribute their skills, experience and judgement to the work process.

We should not therefore allow technical functionality to dominate. User acceptance, usability, user motivation, exploiting human resources and other user needs should play an important part in the further development of manufacturing systems and computer-based technologies.

As new economic and technical forces change the standards for economic competition, they also affect organizational structures, skill requirements and jobs. Organizational formats are shifting towards flexible networks that use information to integrate organizations, expedite strategic changes and improve customer service. In fact, the physical energy necessary to extract resources, manufacture products, and deliver services is becoming less important than the information required to respond to markets quickly. Increasingly, information is becoming the basic raw source for future progress.

9.6 NEW MARKET STANDARDS

Markets are the nexus where producers and consumers come together. A market represents the distillation of human wants and needs into material goods or services. Moreover, markets are a relatively uniform motif in the disjointed pattern of economic change. The basic human wants and needs expressed in markets do not vary much over time: e.g., food and drink, housing, health care, education and training, communication, transportation, entertainment, community, physical and emotional security, and safe and pleasant surroundings. Ultimately, the new economy will be measured by its ability to satisfy these timeless wants and needs efficiently and fairly.

9.7 MARKET STANDARDS–A THUMBNAIL STORY

In 8000 BC humans settled down on farms. Early agricultural production was used most for subsistence or to pay tribute or rents and was rarely sold competitively. Competition and competitive standards were primitive. Subsequently, in 4500 BC, small communities and toll based manufacturing appeared along with early crafts. The labour-intensive craft and agricultural economy developed gradually over the next several hundred years. Competition in agricultural markets accelerated slowly with urbanization, as townspeople created a growing demand for farm supplies.

Output per person remained relatively flat until the eighteenth century. It has been a self starter ever since. Supply and demand have been like the proverbial chicken and egg. Selling of products and services has generated spendable earnings to fuel further expansion. With the aid of productivity increase and invention, expansion continues to include the limits to growth. The spiral or relatively effortless growth has dumbfounded naysayer (pessimist) from Malthus to The Club of Rome, as one doomsday after another has come and gone, and the world still hasn't run out of land nor air.

9.8 THE MASS PRODUCTION ECONOMY: MEETING PRODUCTIVITY STANDARDS

The astonishing productivity growth that began in the 18th century and is continuing today stems from the genius of mass production. In the mass production systems, products and services are reduced to their smallest and most reproducible parts. Machines are then designed to make each individual component. In the stereotypical mass production institution, white-collar and technical elite invented standardized products, design production jobs and machinery and orchestrated the piecemeal output of specialized workers and narrow-purpose machines within carefully organized top-down hierarchies.

Traditional relationships are often adverse and lacking in trust. There is little sharing of information and no working towards a common goal, although many of our enterprises are now working towards improved supplier relationships. Companies are sharing design and production data and building long-term relationship based on trust and collaboration. They are in effect building seamless system, which gives the impression to the outsider that the supplier is part of the enterprise.

In agile manufacturing, the term virtual corporation is used to mean something different. In this situation, efforts are concentrated on building joint ventures with other corporations who have other distinctive and complementary core competencies allowing the creation of several companies and then synthesized into a single entity.

These methods and techniques include just-in-time manufacturing, concurrent engineering, cellular manufacturing, to name but a few. Many of these are useful techniques, but they are often oversold and presented as panaceas. We will have to learn that, while all methods and technique are potentially useful, they are just part of the toolkit of manufacturing. They are not panaceas.

We will have to start asking fundamental questions, not just about what to do, but also about why, when, and how to do these things. In other words, we will have to move away from panacea, prescription and platitude (boredom), towards carefully planned agile manufacturing. This will not be easy to implement considering our tendencies in manufacturing to follow the existing practices.

9.9 WHAT IS THE NEW ECONOMY: HOW IMPORTANT?

A new economy is already upon us. It is pulling threads from the weave of our economic past, creating whole new cloth. The most obvious change in the texture of the fabric is the growing complexity in the pattern of standards that we must meet in order to win in economic competition. In the old economy, nations competed principally on the basis of productivity and prices. Our success as a nation was measured by our ability to produce higher volumes of goods and services with the same or fewer resources. In the new economy, our national competitiveness is based not only on productivity, but also on quality, variety, customization, convenience and timeliness. People are demanding high quality goods and services that are competitively priced, available in a variety of forms, customized to specific needs and conveniently accessible. What's more, people do not want to wait patiently for state-of-the part products and services.

These new market standards have resulted from profound economic and social changes around the world. In the new economy, consumers are richer than in the old economy. They use more time making money than spending it. Today's consumers can afford something better than they used to. They demand new technologies which allow quality, variety, customization, convenience, timeliness and mass production prices. What's more, in the global economy, if American industry doesn't meet these standards, somebody else will.

Central to the new economy are flexible and information-based technologies. In fact, today's most important technology is the computer. In its various disguises, this information-based technology has raised our potential productivity and quality. It has provided sufficient flexibility to tailor products and services to smaller markets and even to individual customers. In addition, by integrating producers and consumers into economic networks, it has helped to create an environment in which goods and services can be delivered globally or locally in a convenient and timely way.

The mass production economy feeds upon itself. The degree of specialization is limited only by the volume of output. Higher volumes justify the cost of specialized machinery and workers. Higher volumes also justify lower prices, which in turn expand market sizes and continues to have its greatest successes there. Craft and service work has been less amenable to mass production techniques. The craft has not disappeared with the evolution of mass production, but continues in areas of economic activity where mass production techniques have yet to penetrate, such as in the skilled trades. Yet the skilled trades coexist uneasily with mass production, that would deskill them if it could find a way.

Productivity is the competitive standard of the mass production economy and goods and services are even more available and cheaper. The Dutch were the world's first productivity leaders, setting the pace beginning in 1870 and the British in the 1890s (Grayson and O'Dell, 1988). The world standard for mass production techniques and productivity has been set since then.

9.10 THE POST-WAR ECONOMY

The productivity performance peaked in America after World War II. The pent-up (built up) demand for customer goods in the post-war era, in combination with manufacturing infrastructure built for war production, generally pushed productivity performance to unprecedented levels beginning in 1946. The economy of the post-war boom made it seem as though nations could produce goods and services on such a grand scale that material wants should eventually be drowned in a set of resources. Our abundant society was both an economic and political miracle. It short-circuited the two toughest questions facing any society: Who gets what? And what do we do first? There was enough for everybody; all that was required was "an equal opportunity" to share in the largesse (generosity) of productivity. Abundance also solved the priority problem to the extent of 57 per cent of its steel, 43 per cent of its electricity and 63 per cent of its oil.

9.11 MEANWHILE IN EUROPE AND JAPAN

As the Europeans and Japanese dug out from under the rubble in the 1950s, their first instinct was to follow the American example. But copying the mass production system proved more difficult than first supposed. The war had taken a fearful toll. Also, the Europeans and Japanese had profound competitive liabilities. Primary among them was the relatively small size of their consumer markets. The scale economies of mass production could be realized fully in the United States without going offshore and competing abroad. In Europe and Japan, however, domestic markets were too small to permit an emphasis on high-volume production of standardized goods for domestic sales alone. As a result, European and Japanese companies were forced to sell abroad.

The complexity of international markets forced the Europeans and Japanese to pay more attention to their diverse customers. German car manufacturers, for instance, had to produce cars not only for Germans but for Swedish and Italians as well. The Swedish market demands cars for harsh winters and rural driving. Fuel efficiency was not a prime concern because gas taxes were low. In Italy, however, the climate was more forgiving and driving more urban, gas taxes were high and registration fees were based on engine sizes. The German car makers learned to produce weather-resistance cars for Swedish consumers and lighter, more fuel efficient cars for the Italian market.

Fragmented markets forced the Europeans and Japanese to focus on flexibility-toward human resources and machine technologies in order to provide a variety of goods and services tailored to market segments. Also, because the Europeans and Japanese could not realize economies from the sheer scale of production, they were forced to adopt more complex competitive strategies and looked for market niches. Rather than confront the United States head-on in the large-scale mass production markets, they took the path of least resistance. The European offered the Volkswagen Beetle instead of big gas-guzzlers. Similarly, mimicking Mac Arthur's strategy of leapfrogging across the Pacific by avoiding major Japanese strongholds, the Japanese decided, as Mac Arthur did, "hit" me where they ain't". The Japanese entered American markets by gaining a toehold in inches neglected by the domestic American producers, often after taking on the Europeans first.

To compensate for their inability to realize American productivity and scale economies, the Europeans and Japanese focused on quality. Also, as they couldn't reduce costs per unit of output simply by increasing volume, they achieved savings by alternative methods: For example, they reduced costs of reworking products by increasing quality in production; they focused on effective work processes, flexible organization designs and superior integration of human and machine capital.

The Europeans and Japanese had to compensate for the natural American advantages on the human side of production. They had fewer, less-qualified workers as they entered the post-war competition. The glut workers, especially as the baby boom entered the workforce in 1960s, allowed employers to substitute unskilled labour for more careful utilization of human capital and a more aggressive focus on learning at school and the job. In addition, some of the technical personnel, home-grown and imported, were of a superior quality and the sheer size of population guaranteed a greater quality of white-collar and technical employees. As a result, the Europeans and Japanese couldn't compete in the development of major innovations. Instead, they focused on the exploitation of new ideas. Rather than expend their resources on inventions, they competed on the basis of their ability to develop new applications quickly.

The Europeans and Japanese also had to compete against the flexibility of American labour markets. In Europe, a strong craft tradition and a powerful left-wing political movement considerably reduced the employers' authority to hire and redesign work. In Japan, employers were forced to provide "lifetime employment" to blunt the more radical policies of a powerful **Marxist labour movement**, especially in industries manufacturing internationally traded products. American employers were able to reduce costs by shedding skilled labour and substituting mass production machinery in combination with unskilled labour. In sharp contrast, the Europeans and Japanese were forced to treat labour as a fixed cost of production and could not easily eliminate expensive skilled labour by substituting them with less skilled labour. As a result, the Europeans and Japanese had powerful incentives to develop and use human capital, whereas American employers were encouraged to rely on special-purpose machines and unskilled labour to drive productivity.

The Europeans and Japanese also found it difficult to match the sheer quality of America's intellectual and financial capital. In relative terms, America has never been a leading saver or investor, yet the United States is so large and wealthy that setting aside even a modest proportion of its Gross National Product (GNP) for investment results in more capital available per worker than in other nations. In the 1950s, for instance, Japanese families saved three times as much as their income as American families, but earned only one-eighth as much. As a result, total investment was twice as high in the United States as in Japan. Similarly, the United States made relatively low investments in intellectual capital, but because of sheer (total) size, fielded the world's largest group of white-collar and technical workers and largest cache (hoard) of basic research resources.

To compensate for America's advantages in the scale of intellectual and financial resources, the Europeans and Japanese tried to make better use of their smallest quantities of resources. They turned to networks both within and outside employer institutions.

As everyone knows, the genie of international trade has been out of the bottle. The combined value of imports and exports is equivalent roughly to a quarter of GNP. The trends towards rationalization are rooted in a variety of facts such as follows:

- Tastes have been homogenized as earnings have equated worldwide; media, marketing and travel have integrated demands.
- Higher incomes have given rise to international markets for nation's specialized products and services, such as Italian textiles, Swiss watches, and Japanese consumer electronics.
- A worldwide reduction in trade barriers began in the 1950 and continued through the 1980s despite painful trade-related dislocation (Doz,1987)
- Advances in communication and transportation technology have allowed multinational companies to serve large, homogenous international markets from the home country, allowed decentralized worldwide production and sales, and reduced the costs for the newly industrializing nationals to enter markets (Vernon, 1987, p.16)
- Organizational experience in American, European, and Japanese multinational corporations allowed rapid expansions when global markets became robust after the mid-1960s.

By the mid-1960s, the latter conditions resulted in the potential for rapid globalization. The spark that ultimately ignited global competition was the need to find new markets for mass-produced goods when existing markets became saturated. Increased productivity in combination with the natural cycles of boom and fluctuations in domestic markets began to create oversupply in the mid-1960s. As domestic markets became saturated, more and more nations began to compete for international customers. Because of increased mass production around the world, the list of basic commodities, products and services that are oversupplied has grown constantly. By the late 1980s, production exceeded demand by about 20 per cent in steel, petrochemicals, semiconductors and cars (Stokes, 1986).

In a perfectly functioning free market, a glut would drive down prices and the least efficient producers would go out of business. But in the modern global economy, there are a variety of forces that inhibit natural demise. Institutions are reluctant to shut down and accept the loss of huge start-up costs. Government support for basic industries can guarantee survival beyond the natural life cycle of economic viability.

9.12 IMPACT OF GLOBALIZATION: GLOBAL STATUS

The impact of the globalization of economic competition has been profound and in some way unexpected. At its simplest level, globalization has increased the intensity and nature of competition. In a world where supply exceeds demand, the competitive importance of productivity and prices is reduced; quality, variety, customization, convenience, and timely delivery of state-of-the-art products and services become the competitive edge. In global markets, it requires increased service functions in order to deliver products and services. In global competition the rule of thumb borrowed from a phrase from the environmentalists, is to **'think globally and act locally'**.

Globalization also seems to have shattered the "product cycle," a self-perpetuating hand me-down process of international economic development that has been in place for time immemorial. Until recently, the logic of the product cycle was historically proven and difficult to challenge. According to this logic, global economic development began in the developed world, principally in the United States, where the markets were the wealthiest and largest, the labour force was the highest paid and most skilled and the financial capital was most readily available. The American market was the legal seedbed for the development of new technologies and products and as a result, the production systems was reduced to simple tricks requiring a small cadre of elite white collar and technical workers who managed relatively unskilled labour and standardized technology. Once simplified, mature production systems were passed on to less developed nations that could use simple technologies in combination with cheaper and less skilled labour. In the meantime, the United States moved on to the next wave of new technologies and products. In this **hand-me-down system**, the developed and underdeveloped nations of the world moved in lockstep up the development ladder. A rising tide in the developed world eventually raised all boats worldwide and did so without disrupting American superiority in the economic pecking order.

This comfortable cycle has broken down under the weight of the new economic reality. Development has made human and financial capital more available outside the United States and markets outside the United States have accrued sufficient wealth to drive new product demand. Further, new global wealth, in combination with new communication, transport, and information technologies, has reduced the scale advantages of the American domestic market. The Japanese market is now almost half the size of the American market and is growing. As economic integration proceeds in Europe, market size buying power and per capita income there and in the United States was becoming roughly equivalent (Dertouzos, Lester, and Solow, 1989). In addition, new technologies and the dismantling of trade barriers had made the size of domestic markets less important. The Japanese, for instance, sold six times as many video cassette recorders (VCRs) around the world as they sold at home.

The most profound assault on the international product cycle had come from a general breakdown in its sequential nature. Today, the nation that develops a new idea may not profit from it before handing it over to a lesser economic power. Technology is footloose. The quality of the indigenous human capital is increasingly worldwide. As a result, nations can step into the product cycle at any given point. Indeed, it is often best to let others bear the cost of development and to focus resources on subsequent phases of the product cycle. There is pride in invention, but there is money in developing products, making small improvements in efficiency and quality and developing new applications for existing technologies, products and services.

Globalization has been a mixed blessing for Americans. In a robust global market, the possibilities for economic expansion are impressive. The potential demand for goods and services in the world economy is vastly greater than current production levels. At the same time, globalization has helped change competitive standards in ways that do not play exclusively to our strengths. Productivity and the price reductions it brings are necessary, but not sufficient, for successful competition in global markets. Our scale advantages are eroding as Europe and Asia are becoming more and more cohesive market spheres. In addition, the Europeans and Japanese have gained more experience. We have to do with flexible production systems necessary to succeed in the highly fragmented global market and to devise a new set of rules and procedures to stabilize world trade. Our domestic markets are no longer large enough to satisfy our productive capabilities and the extension of economic activity into the global market is necessary if we are to continue to increase our own standard of living. Moreover, if we are allowed access to foreign market, we cannot deny others access to ours.

9.13 LIMITS TO GLOBALIZATION OF MANUFACTURING

The new evolution of manufacturing technology (as described so far), in particular, the increase in economies of scale in manufacture has been one of the key conditions in favour of market globalization and MNCs (multinational companies) integration in a number of industries. Several factors may now slow down this trend.

- Firstly, new technology has been so successful at reducing manufacturing unit costs that these costs now account for only a small portion of total delivered costs. Further reduction of manufacturing cost will be of lesser impact that in the past, as other elements of cost play a much greater role, namely, overheads, R&D recovery and distribution.
- Secondly, economies of scale may no longer increase in the same way as in the past. Some new technologies may abruptly decrease economies of scale. New multipurpose smaller processors in the chemical industry are examples of this type of technology. Even in the absence of genuinely new technology that would reduce economies of scale, the advantage of manufacturing systems – from the well-known materials and resource planning systems to the embryonic 'factory-of-the-future' concepts–are based on cost reduction and better management of the manufacturing system rather than from increasing the plant size or the length of the production run. Better manufacturing processes allow more flexibility in production. For instance, multiple car models can be produced in varying proportions on the same assembly line with relatively little cost penalty. This could allow car manufacturers to move back from large single model factories serving single country markets. Although there may still be some cost penalty in setting up a flexible factory (rather than a narrowly focused one), at least the trade-off between increasing flexibility and decreasing costs can be explicitly considered.

The impact of flexible manufacturing on the trade-off between integration and fragmentation of manufacturing is still not clear; however, greater flexibility as in agile manufacturing allows producers to cater to shifts in consumer preference–as their discretionary income increases from cheap standardized goods to customized products. Flexible manufacturing systems may allow product customization – at least so long as such customization can be achieved through factorization around a common core and low cost. These systems may shift the basis for cost advantage from scale to scope and thus make it possible for an integrated manufacturer to serve differentiated worldwide needs.

- Thirdly, the "Just-in-Time" manufacturing concept works best with the allocation of various facilities into an integrated system. This polarizes globalization and integration to the extremes: either a series of small "local-for-local" plants, each by and large self-sufficient, or to the opposite extreme, namely a single integrated source for everything (e.g., Toyota City, or to a lesser extent, Boeing around Seattle, or Caterpillar around Pretoria).

For example, a widely dispersed integrated manufacturing network (such as Ford of Europe) where plants are distant and supply each other with components and subassemblies is least amenable to just-in-time manufacturing.

Some Considerations

In order to meet the impact of globalization of products and manufacturing, it is necessary to consider several factors, such as:

- Supply of customer-oriented products
- Challenges posed by multinational companies, which are already practicing the concept of agile manufacturing
- Need to produce components of international standards, in terms of quality and cost
- Healthy environment and enhancement of skill for the workforce
- Energy conservation and environment-friendly products
- Implementation of green engineering.

The impact of globalization on products and manufacturing education and practice has led to a new educational structure as shown in Fig. 9.1. It can be seen that with effective leadership, integration, internship, operations management, product/process design, manufacturing, TQM, plant visits, etc., it is possible to lay a strong foundation for sustained growth and stability of the new systems all levels.

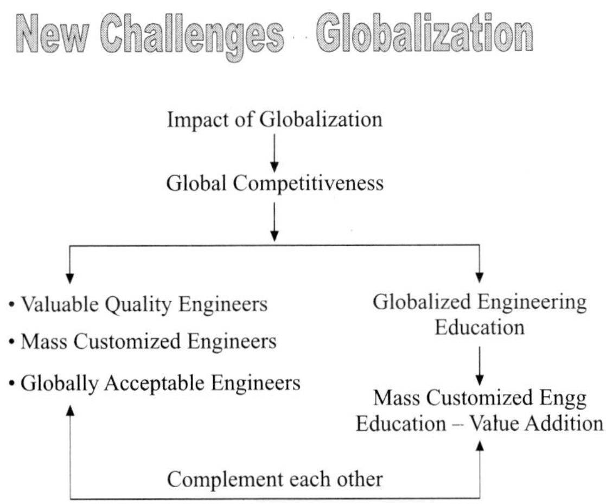

FIGURE 9.1: Impact of globalization: New educational structure.

9.14 THE DIVERSIFICATION OF STATE

Plain vanilla essence wasn't good enough any more. Variety and customization of goods and services had become key competitive principles. Consumers' tastes have also diversified because of a flammable mix of the economies and demography at home and abroad. Increasing economic wealth contributes to diversity in demand in two ways:

- First, it changes what people want. As people get richer, a smaller share of their income goes for the basis of food, clothing and shelter and they begin demanding variety, quality, tailoring, convenience, and state-of-the-art products and services. They also want more intangibles, for example, they can afford to let environmental, health and nutritional concern influence their purchasing decisions.

- Second, growing health gives economic voice to underlying ethics, geographic, cultural, religious and gender differences that were there all along. No nation, with the possible exception of Soviet Union, is more diverse than the United States. Moreover, our diversity in taste is increasing. Demographic changes at home have resulted in a more complex domestic market. The aging baby boom continues to create fresh waves of new demands in its wake. The American family has decreased in size and has increased in diversity. Also, enormous increase in wealth and life expectancy have resulted in new markets to serve older Americans (Kochan, Mitchell, and Dyer, 1982). Globalization of economic activity has also been a major external force in the diversification of tastes.

Eventually, information-based technologies have allowed employers to locate flexibility in the technology itself, as well as in the work processes for using the technology. With flexible software, a few keystrokes at a control board can reprogramme machinery and work processes. The computer has brought a whole new level of built-in flexibility and precision and allows employers to raise the ante (stake) on productivity and quality standards. The computer saves time by allowing workers to make changes with a few keystrokes rather than reconfiguring whole production systems of work processes. The new technology also provides convenience. User-friendly software makes technical complexity as invisible to most customers and workers as the carburettor is to most drivers of cars.

By all reports, the expectations for service are increasing and quite a few, express the growing dissatisfaction with the customer service they get. A conference Board Survey of 6,000 households concluded, for example that Americans were reasonably satisfied with products but pervasively dissatisfied with service quality. The **Yankelvitch Monitor surveyed** 2,500 Americans about their satisfaction with customer service and found that respondents estimated only airline service had improved over time. But a closer look at airline customer service does not accept Yankelovitch's good news. **Zemake reports that, in 1987**, the U.S. Department of Transportation received more than 44,000 complaints from airline passengers, an increase of 25 per cent over 1986. In 1987, only 66 per cent of airline flights arrived on time and airlines lost 11 out of every 1,000 pieces of luggage.

9.15 TIMELINESS

The early bird will win market share in the new economy because time is money. According to one study of high-tech markets, products that come to market on budget but six months late will earn 33 per cent less profit over five years than products that come out on time but are 50 per cent over budget. Institutions compete now in several successive races against the clock:

- First event: Develop a major innovation, whether a technology, a product, or a new work process.
- Second event: Move the innovation off the drawing board and into the hands of consumers.
- Third event: Race up the learning curve to improve the innovation by increasing efficiency, improving quality or developing new applications.

- Final event: Use the knowledge accumulated in the race of the learning curve to make a breakthrough to another major innovation. This race occurs after institutions have wrung all possible incremental improvements and new applications out of the original innovation.

A single employer or nation rarely wins in all these races. In the nineteenth century and first half of the twentieth century, the United States became a world-class economic power by borrowing ideas from other countries. We were not the best in invention, but we were first in the race to get these borrowed ideas off the drawing board and into the hands of customers. After World War II, however, Americans became the wellspring of invention. The United States ended up with the lion's share of the world's intellectual, financial, and physical capital.

Efficiency, quality and productivity. For example, the **Xerox quality programme** reduced production costs by 20 per cent. **Harley Davidson** reduced the cost of reworking defects by 60 per cent by focusing on quality manufacturing techniques. **GM's lansing assembly plant** drove costs down by 21 per cent with embedded quality procedures.

Thus, quality often is the best antidote for a productivity problem. Because most quality problems stem from design. Improved design can mean big productivity gains:

- Experts claim that not more than 20 per cent of quality problems result from production defects. The other 80 per cent occur at the design stage of a product or service.
- More "robust" designs that allow high performance despite production errors have slashed performance defects at **ITT** by more than half and saved more than $60 million.
- **Ford redesigned** an instrument console for the **1987 Escort** using only 6 parts, compared with 22 parts in the original 1984 model. The effect was to reduce material costs by 39 per cent, drive down labour costs by 83 per cent, and improve the defect rate by 10 per cent.

9.16 CUSTOMER FOCUS

In the final analysis, the pursuit of any of the new competitive standards ultimately translates into convenience for the customer. Providing variety and customization begets convenience because they offer choices that meet the specific needs of groups or individual customers. The pursuit of timeliness leads to convenience as well. Employers who try to build speed into their products and services, inevitably end up closer to their customers and these close ties are a fresh source of information on customers' needs and desires. Attention to speed also increases convenience because, for a busy consumer, convenience is largely a matter of time saved. Higher quality products and services are convenient because they work better. Every competitive victory in the new economy results in more convenience for the customer.

9.17 GLOBAL STAKES

Fast growing companies are significantly different from those of slow growing, capital intensive industrial enterprises. Slow growing companies generate most of their capital from internally generated profits. Thus, accelerated depreciation and investment tax credits are important for them. By contrast, fast-growing high technology firms that are not so capital intensive look to the equity market investments in research and development (R&D). Thus, lower the capital gains tax, the R&D tax credits are more attractive than accelerated depreciation to high technology companies.

While the present high cost of capital discourages high risk investments in new technology, the growth and vitality of new industry will be impaired more by a lack of human capital than financial capital. It is significant that the cost of capital is three times greater in the United States than in Japan and much more in India. What is more significant is that in Japan the number of Japanese engineering graduates is three times than the number of American engineers in America on a per capital basis and also more in absolute numbers despite a population of half our size. Technical talent is the raw material that feeds the growth of the high technology industry and we have reason to be concerned about the adequacy of our supply. There is a serious shortage of engineers in America which limits the breadth and depth of product and technology development we can undertake. Underlying this shortage is an under funded and overstretched system of equation.

Neither the strategic importance of education nor its close link to high technology is widely recognized and understood. Characterized by rising costs and falling test scores, our educational establishment has reached a low ebb in public opinion and governmental support is being drastically cut back. This is a **"San Andreas fault"** in Government policy that creates a potential for disaster, as shifting economy and an unsupported system of education will create pressures throughout society. This fault in the national policy of many countries is most visible in the recent phase-down of support for science and engineering education at a time when demand for engineers outstrips supply.

Even more incongruous is the build-up of defence spending, which we cannot normally afford in the face of manpower shortages without a commitment to expand the pool of engineering resources as an integral part of the defence budget. Will the Government's demand for scarce human resources crowd out many of our private sector needs in much the same way that Government demand for financial capital has driven the cost of capital to crippling levels.

The incongruities in national policy, particularly with regard to the role of education, can at least in part be explained by the fact that our society has failed to grasp the full significance of the transition that is now under way. We are moving from a capital intensive, physical -resource-based economy of the first half of this century to a knowledge-intensive, physical-resource-based economy in the last half. The formulas, policies, economics theories and conventional wisdom that facilitated the earlier transition from an agrarian to an industrial society are no longer applicable to the transition now in progress from an industrial society to an information society.

It should come as no surprise that we cling to old ideas and values long past the time when evidence clearly calls for change. Research on changing patterns of history has shown that societal forces tend to reinforce the status quo and even to condemn as heretical (unorthodox), facts and ideas about the world and how it functions. Galileo's prosecution by the authorities of his time serves as a constant reminder that our well established, well intentioned and best-educated contemporaries can be dead wrong about what makes for a better world.

Thomas Kuhn in his seminal work **The Structure of Scientific Revolution** has carefully analyzed and documented the dynamics of change that involves significant transitions of ideas and the values. His study focused on changes in the physical sciences. Later authors extended these same concepts to political, religious, social and economic patterns of change. In the history of science, supposedly the most factual and objective of human endeavours, **Kuhn** points out that established views of how the world works prevail along after a new discovery may have rendered them absolute. For example, **Einstein's theory of relativity and quantum mechanics** revolutionized **Newton's mechanistic** perspective of the world. Yet, at that critical juncture from old to new, nearly all experiments, scientific works and research policies continued to be based on the validity of Newtons laws long after the time when these efforts could yield useful results. Today, we are at a similar crossroads of economic and social history, insisting on using old theories when we already know what works better.

From another perspective, **Jay Forester,** an MIT professor who is building national economic model based on engineering feedback theory, notes that economic vitality waxes and wanes in fifty year cycles the so called Kondratieff long wave. He explains this long-term business cycle by changes in the underlying technological base of the economy and by the tendency toward the end of a fifty-year cycle to over accumulate capital stock in old technologies, which are subsequently rendered obsolete by developments in new technologies.

Whether or not the **Kondratieff** wave presage (foretell) further economic malaise (depression) for the mid and late 1980s, Forester makes a significant point that there are long lags between the time, when capital investments are made and when the results and payback from the investments can be measured. During periods of transition, this can lead to an "overshoot" of investments that are written off.

When juxtaposed with the observations of **Kuhn** and others that societies tend to continue their most recent behaviours pattern, regardless of evidence to the contrary, **Forester's** theory takes on adding meaning.

Robert Hayes, a professor at the Harvard Business School, also challenges today's conventional wisdom. The usual reasons given for declaiming productivity, he says, do not hold up under close scrutiny. Such impediments as insufficient capital investment, high energy costs, labour problems and Government red tape may contribute much to our problem but little to our thinking about its solution. He observes, for example, that France and Germany have these same problems and yet the rate of their productivity growth has been significantly higher than ours.

Hayes concludes that meaningful strides in productivity do not come so much from incremental investments to improve existing products and manufacturing processes as from investments that create products and markets based on substantially new different technologies. What is wrong, Hayes argues, is the preoccupation by management with short term incremental investments in existing businesses to the detriment of long-term investments in new technology and market development. By this reasoning, it is technological innovation more than capital formation that will solve our productivity problems and innovation from investments in people as well as investments in machinery.

From many perspectives, it can be argued that more of the same is not what is needed. The longer we resist and prolong the transition, the more radical and painful the adjustment will be, to a new economy. To those in high tech industry, the transition has already taken place and it is a fact of life. But among the wider public and most decision-makers, the realities of a new era have not yet been fully understood or accepted.

The fact that high technology embodies an unprecedented amount of human knowledge and technically sophisticated labour will change the equations by which national priorities are calculated. This is not to deny the continued importance of natural resources and the need to conserve non-renewable ones more wisely. Nor is it to repudiate (reject) the role of capital and the need to control inflation more vigorously. But once the concept is fully grasped that knowledge should be seen as a strategic resource with an importance equal to or exceeding natural resources and physical investments, then a chain of propositions follows that will change the way of national priorities and strategies.

10

Agile Manufacturing: Implies Revolutionary Changes

There is increasing awareness of the importance of issues such as organization, people, design process, and so on. However, increased awareness has not always been accompanied by fundamental changes in values, beliefs and practices. The established paradigm has in many cases remained largely unaltered. On the whole, we have often chosen to modernize the traditional manufacturing paradigm, rather than to replace it completely with new paradigms.

Organization and people's issues illustrate very well the problem we experience. While most of us now accept that these matters need to be considered in the design and implementation of manufacturing systems, we still adopt a fairly narrow orientation to how these issues should be considered. Our orientation is often restricted to forcing the organization to adopt itself to a new technology, often to a great cost, overcoming the resistance to change. We commonly assume that resistance is derived from the reluctance of people to accept new technology. We rarely perceive that resistance might arise from a mismatch between the organizational, human and technological elements of manufacturing systems.

The fundamental changes that are now occurring in market conditions not only demand new techniques, but also a fundamental shift in paradigm. Paradigm shifts are however, difficult to achieve and may be opposed to preserve the existing paradigm.

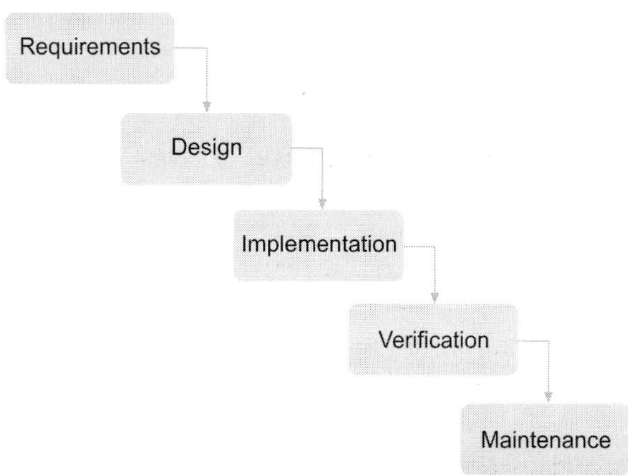

FIGURE 10.1: Waterfall life cycle model (software development).

In the field of software development, the traditional waterfall life cycle model, shown in Fig. 10.1 is largely obsolete, except perhaps for a few specific classes of software projects. The problems associated with the use of this approach are now well documented and given the high-risk nature of market conditions and the costs of failing to produce systems that meet the needs of customers and users, this stage-wise document driven process should be replaced more by the spiral life cycle model shown in Fig. 10.2. This approach is a risk driven evolutionary approach, where we place our emphasis on reducing identified risks.

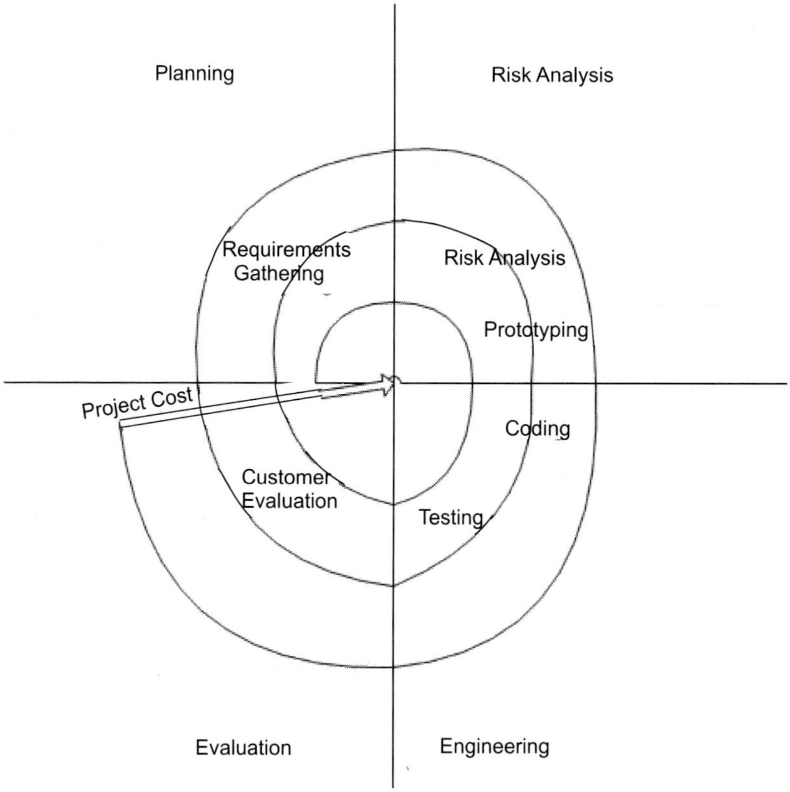

FIGURE 10.2: Spiral life cycle model (software development).

In the coming years, we will see a growing emphasis on what has been called idea transformation (Savage, 1992). During the industrial era, material transformation has been a dominant consideration, i.e., the process of transforming materials into products. This is depicted in Fig. 10.3.

The changes will also be painful for us. Savage (1992) refers the pain aspects by referring to discontinuous brief patch that lies between our traditional paradigm and the new paradigm of agile manufacturing.

Some companies have already started to change the way they use technology and have begun to consider organization, people and other issues. The changes needed are simple in principle, but in practice they can be difficult to achieve. Instead of looking for the technological fix, we need a broader approach that addresses organization, people, technology, product design processes, relationships with suppliers, and so on. We have to overcome our belief that technology provides the answer to all problems. We should not use technology to compensate for our shortcomings.

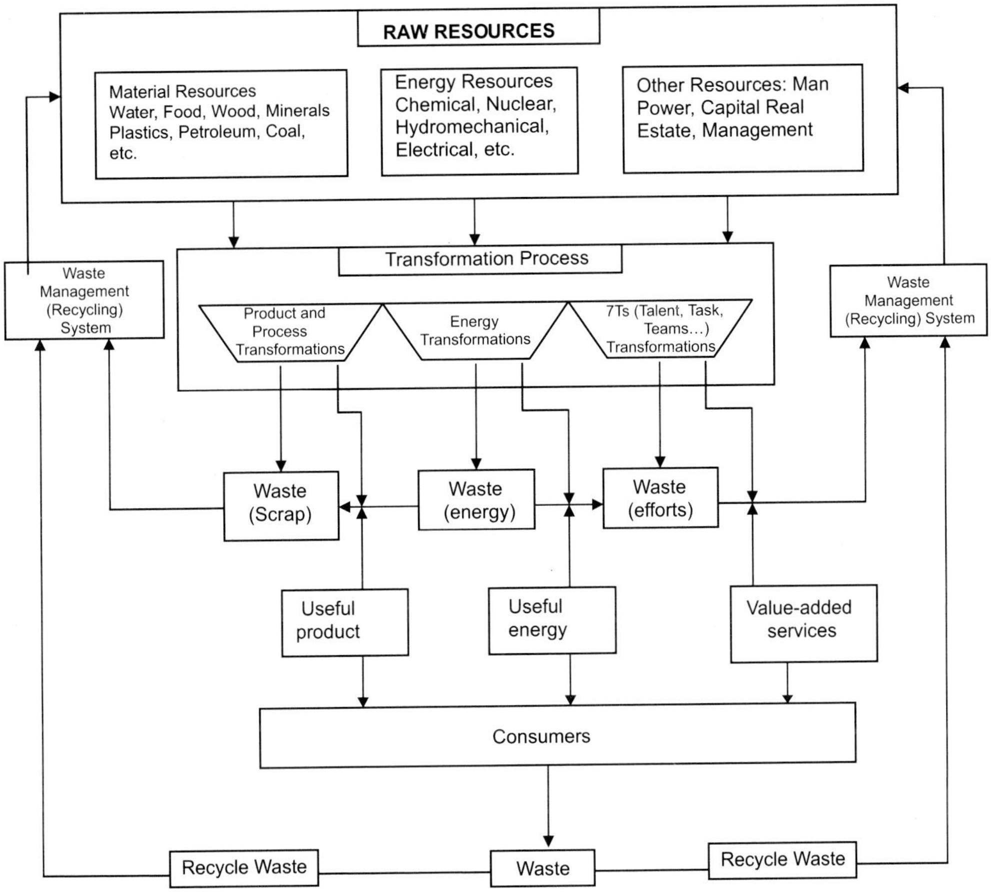

FIGURE 10.3: Materials transformation: Process of transforming materials into products.

10.1 THE CHALLENGES OF CHANGE

Integration of organization, people and technology into a coordinated and interdependent system will be a new experience for all of us. In order to achieve agile manufacturing, we will have to recognize that we are pursuing a moving target, because the state-of-the-art will always be advancing. Along the road, we will be faced with hundreds of methods and techniques. We should understand the wide ranging nature of the changes that lie ahead and also that we have to change. Change is not just something that others have to go through.

10.2 EARLY STEPS AND PREREQUISITES: STANDARDIZATION

This presents the prerequisites to develop mass customized products and early steps the companies can implement to prepare them for subsequent mass customization activities. Since standardization helps all product developments, it should be implemented right away even if a company has not yet fully committed to mass customization. The methodologies presented here are not capital intensive and can be easily implemented without many levels of approvals. Further, these steps can be justified on their own merit.

Standardization of parts is also extremely important to ensure that flexibility will not be compromised by the setup of retrieving parts, changing dies, changing programme or "doubtful loading" automation machines, which are necessary if there are more parts or tools than bins of receptacles; the machines will then have to be run twice, once to assemble half the parts and again to assemble the remaining half.

The issues that affect low-volume products are more subtle but equally important.

Every company has similar horror stories regarding part proliferation. One might ask why, especially considering the following facts:

- Part proliferation impedes manufacturing flexibility and thus, discourages mass customization
- Most part proliferation is unnecessary
- Part proliferation generates a great deal of overhead cost

10.3 THE COST OF PART PROLIFERATION (Ref: Section 10.12 also)

Part proliferation is expensive. A **Tektronix study** determined that half of all overhead costs related in some way to the number of different parts handled. Most companies do not even know how much this cost in dollars. A **survey of several Fortune 500 manufacturing** companies revealed that not a single company or division had an accurate estimate of the cost of a part over its lifetime! According to **Venkat Mohan of Cadis, inc.,** who 'intuitive estimates range from $100,000 per part for custom parts. **"James Shepherded, Director of research for Advanced Manufacturing Research (AMR), Boston,** says that in electronics, the cost of just entering new purchased components is between $5000 and $10,000 per component. The **Ernst and Young Guide** to Total Cost Management states, "It is not surprising that manufacturers have estimated the annual administrative cost of each part number to be $10,000 or more.

In addition to those official materials' costs, excessive part proliferation adds cost of service and manufacturing cost, but rarely related to cost for setup, inventory, floor space, lower machinery utilization and other flexibility issues.

Part proliferation also lowers assembly productivity. Writing about the automobile industry in a **Wharton Business School report, Fisher, Jain and MacDuffie** write, "Part variety also appears to have the greatest negative impact on assembly plant productivity".

When a large machine tool company investigates this phenomenon, it discovered it had 521 very similar gears. The company eventually reclassified all those gears into 30 standard gears. There were nearly 17 times more gear types than necessary. Every gear had an average of 17 functional or exact "duplicates".

10.4 MORE FLEXIBILITY/AGILITY

Eliminating setup changes allows products to be built in any size lot. With setup reduced to zero, any quantity of any product in the family can be built. This flexibility is the key to building products to order that allows companies to mass customized orders.

10.5 BUILD TO ORDER (Ref Also. Section 1.1)

Flexible manufacturing in turn, can eliminate finished goods inventory and let the plant build only the products that will ship immediately (for which it has orders). This is knows as "build to order". Another important benefit of build-to-order manufacturing is that scarce parts are only consumed in products that go immediately to customers. Further, expensive parts are consumed closer to the time of the same, thus lowering interest expenses.

10.6 WHY REDUCING INTERNAL VARIETY IS SO IMPORTANT?

Implementing commonality for parts, features, tools and materials can benefit any company, even before it embarks on mass customization. Since commonality is a prerequisite for mass customization, commonality programme is one of the first steps to implement. Since commonality generally applies only to new products, the programme should be implemented as soon as possible in order to reap the benefits at the earliest. The following discussion presents the benefits of commonality under four categories that are very important to mass customization: flexibility, responsiveness, cost reduction and quality.

10.7 ELIMINATING SETUP

If the number of parts used in manufacturing is small enough, those (commonality) parts can be permanently loaded on assembly machines or in the manual assembly pick bins. This allows all the products to be built by the same process without having to change the setup for different parts.

10.8 INVENTORY REDUCTION

Inventory can be significantly reduced with fewer types of parts to stock and distribute. This will reduce incoming or "raw" parts inventory. Parts commonality and setup elimination also encourage just-in-time operations that can drastically reduce work-in-process inventory expense and floor space.

10.9 INTERNAL MATERIAL LOGISTICS

The flow of parts within the plant will improve with fewer parts to order, receive, log in, stock, issue, load, assemble, test, and record. Having few enough parts to distribute at their points of use will avoid the space and expense of kitting parts.

10.10 BREAK STRUCK DELIVERIES

High-usage, low-cost common parts can be delivered right to the point of use without any overhead costs for purchasing or internal distribution. Arrangements can be made with suppliers of low-cost parts who will simply keep the bins full of common parts at all the points of use, much like "break struck" deliveries keep the shelves stocked with break in markets. The supplier simply bills the company for each month's usage. This is a very attractive approach for low-cost common parts such as fasteners, washers, resistors, and so on.

10.11 REDUCING INTERNAL COMPLEXITY

An important prerequisite to flexible manufacturing and thus, to mass customization, is the reduction of internal complexity, which requires standardization efforts for parts, materials, tools and processes.

10.12 PART PROLIFERATION (Ref. Also Section 10.3)

Recent competitive pressures and currency issues are forcing Japanese automobile manufacturers to reduce cost. One of their thrusts is to investigate and reduce the cost of internal variety. As a result of such an investigation, **Nissan** realized in **1993** that, in the current model line-up, it used 110 different radiators, 300 different ashtrays, 437 dashboard meters, 1200 floor carpet types and 6000 different fasteners. This type of variety does not add value to the customer. Of course, customers want their carpets to match interior fabrics and exterior plant but this could be accomplished with less carpets type.

An electronics company had 1,500 types of resistors, including 120 different kinds, size and tolerances of 1,000 ohm resistors. That company was able to reduce the 15,00 resistor types to less than 200 actively used for manufacturing.

Another client, who made consumer products, discovered he was using the following numbers of part types: 1,248 wire assemblies, 152 motors, 151 screws, 74 switches, 67 relays, 65 capacitors, 37 valves, 16 transformers, 62 types of tape and 1399 different "standard" labels.

Thus, the challenge for mass customizers is to reduce the internal variety to the point where products can be built flexibly without much increase in cost and time delays of setup changes. This is accomplished by standardizing parts, features, materials, and processes in the design stage.

Flexible manufacturing is a prerequisite for mass customization. Issues like setup reduction have obvious implications to high-volume producers and should be fully discussed. To operate flexibly in high volume, parts must be common enough to be readily available at all the points of use; fixture geometry must be common enough so that all variations of parts can be positioned, design features must be common enough to avoid stopping production to change materials. If parts cannot be distributed at their points of use, then they will have to be "kitted' together in batches with enough parts for the number of products to be built. Kitting inhibits flexibility and discourages mass customization.

All this incurs significant overhead that is more proportional to variety than to volume, especially for parts that are small and do not have special requirements for inspection, handling or storage.

Consider the difference in material overhead between 100 different resistors types (each with a consumption of 100 a month) and 10 resistor types (each with a consumption of 1000 per month). Either scenario represents the consumption of 10,000 resistors per month, but the overhead cost to produce and distribute 100 types would be an order of magnitude greater than for the 10 type.

Mass customization means being able to build, in any volume, products that are customized to individuals or market niches at mass production cost and speed. The formidable challenges with mass customization is to produce an almost infinite external product variety incurring the excessive cost and delays of internal variety that would be inevitable if the typical mass production company attempted to customize product.

10.13 ENCOURAGING STANDARDIZATION (Ref. Also Section 10.2)

Given the importance of internal variety reduction to mass customization, it is imperative that mass customizers or potential mass customizers encourage standardization implementation as early as possible. This means encouraging design engineers to design around common parts (even ones that might appear too expensive), specify common design features, designate common tools and-base designs on common materials. In fact, engineers may be actually pushed the other way by poorly conceived metrics, such as emphasizing parts' cost and low bids for single products instead of total costs for product families.

The following steps can be taken to encourage standardization. They involve "discounting" material overhead rates for common parts, pre-qualifying common parts, making samples and specifications of common parts readily available and emphasizing total-cost thinking, preferably incorporated into total cost accounting to emphasize in goals, policies, directives, "pep talks" and training the importance of commonality.

10.14 PROCESS STANDARDIZATION

Standardization of processes results from concurrent engineering of products and processes to ensure that the processes are actually specified by the design team, rather than left to chance or "to be terminated later". Processes must be coordinated and common enough to ensure that all parts and products in the mass customization family can be built without the setup changes that would undermine flexible manufacturing.

One of the processes affected by part standardization is mechanized screw fastening. Auto feed screwdrivers are very cost-effective mechanized tools that orient and feed the screws which are then blown down a tube to the screwdriver head, where they wait for the operator to activate the power drive by pushing down on the screwdriver handle. A present production tool can feed any style of screw (machine screws, self taping screws, etc.), but only one size and one type at a time.

10.15 MATERIALS COMMONALITY

Mass customization needs raw material commonality for products whose customization is accomplished by "cut-to-fit" operations. If the raw materials to be processed can be standardized, then the processes can be flexible enough to make different products without any setup to change materials, fixturing mechanism or cutting tools.

Thus, lack of standardization, in itself, could be a competitive liability to low-volume producers because it would cause a relative disadvantage with respect to cost, delivery, quality, repair, field service, and possible reputation.

11
Agile Manufacturing: Advantages and Applications

11.1 ADVANTAGES OF AGILE MANUFACTURING

- The first is the change from standardization to post-standardization. Industrial society was characterized by high-volume production of a small variety of items. An information society, however, is capable of small production runs of multiple items which match individual needs.
- The second is a shift from specialization to collaboration, with many aspects of the society becoming increasingly composite and multidisciplinary.
- The third is the shift from the constraints of time and place.
- The fourth is the trend away from concentration towards diversification.
- The fifth is the movement from maximization to appropriate scale (from big is best, we are moving toward 'small is beautiful') finally,
- The sixth is the trend away from centralization toward decentralization.
- Finally, the seventh is the lead to sustainability in manufacturing.

These changes will bring shifts in the pattern of industrial activity. For example, post-standardization, as it affects the distribution sector, will bring about diversification in retailing, such as non-store retailing, home shopping and the like. To take another example, that of collaboration, integration of distribution of goods and of finance with the information industry is already occurring.

These transformations will trigger a restructuring of industry. The traditional industrial classifications of primary, secondary and tertiary sectors will become increasingly blurred through the emergency of C and C, the hardware of secondary industry will merge with tertiary industry's software. The result is what might be called 2.5 industry. An example of a 2.5 industrial product is an automobile containing microprocessors. This allows the driver to programme various desired driving conditions which the automobile will proceed to execute flawlessly. Computer-aided manufacturing is an example of a 2.5 industry manufacturing process. With no drop in efficiency, different products can be produced simply by making changes in the software in short, a flexible manufacturing system. This demonstrates the integration of hardware and software taking place in both processes as technology integration in agile manufacturing.

Clearly, materials technology and energy technology will always have an important place in any industrialized society. But the emphasis in the overall development of industrial technology has shifted in the 1970s away from an extremely energy-intensive and materials-intensive style of mass and flow production to a much more flexible information-intensive style, associated with computerized control and communication system.

11.2 APPLICATIONS OF AGILE MANUFACTURING

11.2.1 Prominent Examples

There is a movement afoot that is changing forever the way we create, produce and sell products and services. Although some parameters are still being defined and some support systems are still being perfected, there is now a consensus as to what to call this latest in a seemingly endless succession of Business and Management 'revolution'. The next big competitive wave will be creating the Agile Virtual Enterprise.

Like such highly touted predecessors as **Total Quality Management (TQM)**, lean production and reengineering, agility promises all sorts of wondrous things: faster-to-market times at lower cost, great product customization with shorter cycle times, smarter and happier workers, increased quality and productivity, satisfied customers and a healthy bottom line. Whether it actually delivers these benefits or not depends on how ready the organization is to accept and implement the agility concepts and techniques.

As of now, proven, the process called **Continuous Transitioning System (CTS)** has been implemented successfully in hundreds of organizations of all sizes in virtually all industries.

For the change-dazed manager, the news that there is yet another paradigm shift in the making can hardly be an occasion for joy. After years of being bombarded by one new management or business concept after another, one might be forgiven for wondering 'when does it all end'? The short answer is–it does not. Change is now a permanent part of the global business landscape. The business that will survive and prosper, even in the near term, are those that are lean, flexible and yes, agile, enough to thrive on change and respond quickly to ever more unpredictable circumstances. Inflexible companies that are slow to market and are habitually caught by competitive surprises will find themselves forced to operate in a reactive catch-up mode. Operating in a situation of constant crises eats away profits, produces mistakes and ultimately brings about the demise of the organization.

Other economic indicators paint an equally bleak and disturbing picture. Since 1990, the average annual in the Gross National Productivity of the Unites States, though roughly equal to that of the 12 European community countries, has lagged significantly behind both Japan and the 'Little Dragons' (Hong Kong, Singapore, South Korea and Taiwan). Perhaps more stunning was the rapid decline of the merchandise and manufacturing trade balances of the United States. Both these negative balances were around $150 billion annually.

This reality has galvanized U.S. industry into action, and it now leads in absolute productivity. The agile paradigm holds great promise for companies that have dedicated to realizing their potential. More than anything else, agility connotes the agility to prosper in a time of uncertainty and unpredictable change. As an **Agile Manufacturing Enterprise Forum (AMEF)** report puts it:

'the companies that thrive at integrating continuous change into the day-to-day activities of every employee will be **AGILE.** They will produce:

- The right product
- At the right place
- At the right time
- For the right customer
- At the right price

If these answers sound familiar, it is because the agile enterprise is not so much a 'new' idea as it is a fresh synthesis of the hundreds of ideas about speed, continuous improvement, quality and customer satisfaction that have evolved during the past decade. Indeed, one of the management challenges in moving toward an agile environment will be to convince workers that this is not simply the next 'flavour-of-the-month,' but the logical extension of existing change programmes.

Though there are different 'spins' being put on the definition of agility, there are specific elements that are generally agreed upon. In an agile enterprise, products will be built quickly and at a lower cost for a customer because of detailed data received at the point of sale. This system will provide thousands of choices and options, but customers won't perceive it that way.

With agile businesses, the mass methods of producing a wide variety of standardized products that sit around waiting for someone to buy them will become expensive anachronisms. In an agile environment, there will be simultaneous mass production, modest production, and one-of-a-kind production, depending on the product, market, and customer.

The structural and organization requirements of agile enterprise include

- A flat, fast, flexible organization, with continuous interaction, support, and communication among various disciplines, and no fat in the middle;
- Highly decentralized management that recognizes what is knowledge base is and now it can manage that base most effectively;
- A multiskilled, flexible labour force in which blue-collar workers will more closely resemble white-collar workers in activity and responsibility;
- Extremely flexible strategies and relationships with suppliers and even with direct competitors (these arrangements allow partnerships to be formed rapidly–both within and outside the core sphere of expertise or business-end disbanded, when new opportunities arise; i.e., 'virtual' enterprises);
- Complete mastery of lean production and concurrent engineering;
- Flexible tooling and automation;
- A deep understanding of the customer-based operating on integration and real-time market data collection and analysis;
- And most important a complete mastery of the tools of continuous transitioning.

The agile movement is being led by an organization called the **Agility Forum**, which is affiliated with the **Iacocca Institute at Lehigh University.**

11.2.2 Scenarios Galore

Some of those scenarios provide an awe-inspiring glimpse of what competition will be like in the not very distant future. Consider, for example, the vision of the **'three-day car'**. In this scenario, a customer goes to his or her dealer and helps design the vehicle he or she wants, selecting from a range of options on a computer workstation. Within certain limits, features such as colour, interior fabrics, seat contours and instrument panel layout might be custom designed. After checking to ensure that the selected configuration is feasible, a distributed **CAD/CAM system** then enters the data into the manufacturer's production system, which automatically relays the appropriate information to supplier plants. Components are built to order and shipped to the assembly plant in the proper sequence. At the bottom line, the customer receives a car built to his or her specifications and deliver within three days as compared to the six-week lead time that U.S. automakers typically require to produce a vehicle to order. Does that sound far-fetched? **General Motors Corp** is already gearing up three **Michigan** assembly plants to fulfil the promise of this scenario within the next three to five years.

Not only does the agile approach produce happier customers and growth, it also saves money. Long production lead times are a major reason why auto dealers stock hundreds of cars on their lots; this leads to heavy inventory carrying costs. If a custom-built car from the factory had a short lead time and required no dealer stock, the resulting savings would offset the higher materials cost of a modular structure, custom built car and increase profit to boost.

At first blush, all this sounds an awful lot like what is now called the 'lean' or 'flexible' approach to running a business. There are similarities, but there are fundamental similarities as well. For one thing, lean production is at heart simply an enhancement of mass-production methods, whereas agility implies breaking out of the mass production mould into mass customization, creating an entirely different organization that can produce highly individualized products and get them to market in half the time and at half the cost.

Agility focuses on economies of scope, rather than economies of scale, ideally serving ever-smaller niche markets, even quantities of one, without the high cost traditionally associated with customization. A key element of agility is an enterprise-wide view, whereas lean production is usually associated with the efficient use of resources on the operations. Floor agility also embodies such 'social' concepts as self-directed business cells and virtual partnerships for the rapid formation of multi-company alliances to introduce new products to the market in ways previously considered impossible.

11.2.3 Architecture of Agility

The different challenges facing businesses today require organizations to be transitioned into flexible, agile structures that can respond to new market opportunities with a minimum of new investment and risk. Thus, we must eliminate the 'walls' that traditionally segregate the resources of an organization along purely functional lines and remove the barriers that impede high performance. Management must trust employees to work within the organizational framework, rather than micromanage. Ultimately transitioning requires creation of a whole new organization.

What will the agile organization look like? For one good example, we might look to **Hollywood**. Since the demise of the studio system, virtually all movies have been made by teams of people brought together on the basis and lines up a director who, in turn, usually helps pick the stars. Then a crew of agile people–set designers, cameraman, makeup artists, lighting experts, film editor, etc. brought together as a virtual disbanded team and the people go their separate ways to pursue new virtual partnerships.

The agile organization will be a concurrent enterprise where everything will be performed faster and cheaper and with higher quality, because people share information and do things simultaneously. As customers, we all know the importance of finding and romancing a good contact person in our supplier organizations who will give us the real story. That will not be necessary in the future. Customers will be fully integrated into supplier organizations throughout the entire product realization process from call to cash, without having to negotiate any department walls, because there would not be any.

Roger Nagel of the Iacocca Institute has identified four strategic principles of agility:

- Agile competition in large and small organizations will be based on the agility to thrive on change and uncertainty. Use an entrepreneurial organization strategy.
- In an agile environment, people's skills and information are the most important organizational assets. Invest to increase the strategic impact of people and information on the bottom line.
- In an agile world, cooperation enhances competitive capability. Use the virtual organization strategy as a dynamic structure both inside and outside the enterprise.
- Agile customers pay a fee for skills, material and get a modest profit for products; or they pay a percentage of the perceived value for solutions. Adopt a value-based strategy to configure your products and services into solutions for your customers.

In short, an agile company has customer-integrated tightly linked with a flexible organizational structure keyed to rapid time-to-market for new, high-value, high-quality, easily customized products.

Agility requires the linkage of technologies to organizational structures that can fully exploit the possibilities for profit in sudden market opportunities. Companies that do not develop the capacity to respond rapidly to customer demands will cease to exist.

Companies that provide products and services to families often keep tabs on their changing needs through fast-food chain birthday clubs, which alert them to major changes in a family's buying habits. They know when children will be attending school, when they will be learning to drive and when the parents might need a home-equity loan to pay college tuition bills.

The life cycle of today's products is only one-third of what it used to be. There are far more options and much greater pressure to reduce size and to customize products for ever-changing needs. **Sony Corp** produces a thousand new products a year, and still customers says 'Could you make it just a little bit different for me?' Times have changed. You cannot survive by making products and putting them in your warehouse. Sell your warehouses! Close your stockrooms! It's production line to customer now.

The most difficult new reality for many managers to understand and deal with is the fact that customers are now looking beyond product and service catalogues when selecting suppliers. They are paying much more attention to the core competencies and the skill and knowledge that the company and its partners have built up over time. Thus, products and services are not the only things they should be selling. To put it even stronger than that, the only things to sell are the skills, knowledge and information of people and partners.

Success depends on the ability to know the customers, to understand what they want and to produce it in a timely fashion. Learning how core competencies and skills can be leveraged to make a difference for customers will have a positive effect on business.

11.2.4 Workforce as Major Asset

In the agile environment, the ability of a company to unlock its people's potential is critical to its success. Look at **Southwest Airlines Co.** Its employees are proud of their company and their jobs. The chief competitive weapon of **Southwest Airlines** is its people. Its competitors have the same equipment, can fly the same routes and many even have better processes or technologies. It must be unique and not readily duplicated. Processes, advanced technology and strategy can be copied quickly but people cannot be. Knowledge capital is little understood as a key to success. Acquiring, developing, retaining and moving the knowledge capital to the 'moments of truth'–the customer contact points will sustain a competitive advantage.

Essential Conditions for Agility Empowerment

At least six out of the following eight conditions must exist for empowerment to take root in an agile organization:

(i) Employees must be properly trained
(ii) There needs to be a shared vision
(iii) There needs to be a whole new set of values for rapid culture change
(iv) The reward system needs to be structured
(v) Managers must trust employees
(vi) The overall culture of the organization must support risk-taking
(vii) Lack of focus on business plan objectives
(viii) Not managed as part of the business

In the course of their exhaustive research into the structure of world-class organization, **Craig Giffi, Aleda V. Rotj, and Gregory M.Seal, authors of 'Competing in World-class Manufacturing: America's 21st Century Challenge (Business One Irwin, 1990)'** uncovered a single yet profoundly elegant principle: all world-class competitors–regardless of industry or size–operate within a common integrated framework. The enterprise of the integrated framework is a total commitment of the organization to product quality and customer satisfaction.

In an agile environment where flexibility and change are the only constants, quality is directly linked to continuous process transitioning as product improvement. To a lesser extent, it often-brings career-limiting punishment. They try to develop the perfect solution, weighting the risks, analyzing the options and calculating the best plan of attack.

One has to put his faith in employees. When empowered they will inevitably do the right thing. In either case, their action is usually better than management's because they are closer to the situation. When people are punished for mistakes, they will not take risks. Mistakes will be hidden and will not be viewed.

One of the best examples of a risk-taking culture is **3 M**. For many years, the stated company's goal was that 25 per cent of its revenues come from products introduced in the five years. In a previous year, the bar was raised to 30 per cent of revenues from products introduced in the past four years.

Any manager or employee who comes up with a new product or idea is encouraged to lead the product development and marketing effort for that product as if it were his or her own business. **3M** sets up the new venture as a separate business from the start. The project manager remains in charge until the project becomes a full-fledged business or until it needs research, marketing, manufacturing or finance.

If the start-up business is successful, the 'entrepreneur' remains on top, with the rank, compensation, bonuses and options appropriate to the level. The company's commitment is limited to success.

By official policy, all **3 M** technical personnel have the right to claim 15 per cent of their time for projects of their own choosing. In reality, some people spend none of their time on their 15 per cent, whereas others spend 100 per cent of their time on their 15 per cent. The strength of the programme is that nobody keeps score. The rule is not so much a rule as it is a permission to be creative.

3 M considers failure to be a step on the road to eventual success. Some people say that the company's **"11th commandment" is 'Thou shall not kill an idea'.** That's the kind of thinking we need to build an agile enterprise.

11.2.5 Technology Revolution

Scientific and technology revolutions of recent years have again changed the landscape of modern industrial life. Production methods have changed beyond recognition, entirely new materials and technologies are being introduced every day and the labour content of products has fallen dramatically. We are far more dependent upon machines and plants than ever before. The rate of change of technology, with its resultant 'future shock' has transformed our approach to working life. A survey concluded that the average worker in the United States needs to be retrained every five to seven years, rendering traditional skills a disadvantage because they must be unlearned. The rate of change is rapidly increasing and the major challenge we face is our ability to harness that change and adapt to it.

Throughout the 1950s the United States dominated manufacturing industry. The economics and manufacturing industries of Japan and Western Europe had been destroyed in World War II. The aftermath also initiated social changes in European countries as ideas of democratic socialism were put into practice. Government regulation on ownership of industrial enterprise, nationwide welfare and health-care systems, government sponsorship of industrial research and education and the rise of labour movements introduced cultural changes that significantly impacted industrial life. The Unites States found itself on top of the industrial heap. Its rich resources, educated workforce, superior technology and vibrant manufacturing industry far outstripped the rest of the world. From the early 1950s through the late 1970s, America was successful in every way—from new products to productivity, from research and development to exports. No one could touch US industry.

12

Agile Manufacturing: Needs Many Inputs

A competitive manufacturing system delivers high quality customer desired products faster than competitive and demanding customers. To do this, however, it is necessary that the manufacturing system is computer controlled and flexible with part flow being efficiency scheduled. Also it is important to treat issues such as product development and product delivery as horizontal business processes that cut across functional departments. Strategic alliance and networking are also vital for meeting the customer needs. An overview of current developments in these areas is detailed below.

Throughout history, manufacturing is going through successive periods of great changes. New materials, new techniques, new technologies have always been at the root of these changes. Currently globalizing markets, demanding customers, congenial environment for foreign investment and other factors such as these have made "manufacturing" more complex. In the current environment, manufacturing has to be interpreted as consisting of a chain of activities. There is need to conceive, develop, manufacture, distribute, sell, maintain and dispose the products while following regulations on environment and keeping internal and external customers satisfied. Technology, flexibility, process management, strategic alliances, supply chain management are important to meet these objectives of manufacturing.

Advances made in the field of science and technology inevitably culminate in manufacturing products that would serve mankind. Manufacturing, therefore, is facing challenges like never before. As new materials emerge into the world to meet the stringent requirements of performing functions for which they are selected, newer and appropriate manufacturing processes must be thought of. Mankind has been witnessing these changes with a hope that it would drive benefit of the advances. Big strides which information technology has been making, has greatly improved information technology's contribution.

Today manufacturing does not just mean mere technology, but much more than this. It is essential that the business of manufacturing should encompass many more components in a cost-effective manner. Then the business houses can compete in the world market with products of International standards beating the time. Capability is not just the criterion but how quickly this capability can be brought forth fruitfully is the criterion in today's world. Manufacturing community looks forward to share knowledge for mutual benefit despite political overtures.

Agile manufacturing, yet an infant in the major part of the world, provides the flexibility, the manufacturer needs. Liberalization has brought into the Indian scene a sense of discipline and competitiveness. It is heartening that an academic institution like the Bangalore University has taken up the responsibility of exposing the industry houses to the concept and advantages of Agile manufacturing.

Agile manufacturing is the science of a business system that integrates management, technology and workforce, making the system flexible enough for manufacturers to switch over from production of one component to another component in a cost-effective manner within the framework of the system.

Keeping the various facets of agile manufacturing in view, a few sub-themes and major aspects under their purview that deserve attention were included for interaction among the experts in the field. The sub-themes include Competitive Foundation, Enterprise Characteristics, Manufacturing Enterprises Elements and Implied Enabling Subsystems. Looking at the infancy of the subject of agile manufacturing, in general, at the international level and 'yet unheard of state' in the Indian context, the response of experts from presenting their work at the Bangalore conference has been impressive.

Manufacturing science has become an inevitable part of the modern life. Meeting specific human needs has become increasingly complex and has now become both inter- and multi-disciplinary subject. The customer requirement that is evident from the present market scenario stresses that manufacturing science needs to be agile enough and capable of meeting even one-customer market. This new trend has resulted in the birth of a new concept in manufacturing, with the main objective to facilitate the manufacturer to switch over from one product in production to another product at a very short notice.

In order to effectively implement the agile manufacturing concept, intelligent integration of technology management and workforce becomes essential. This, in turn, needs application of information management through networked data bases. The advancements in the field of information technology have made an effective integration possible, especially, from the point of view of applicability of existing software and development of new software serving various sections of the manufacturing arena. An efficient integration of need-based software for different applications is, therefore, essential to the successful implementation of agile manufacturing in industry houses.

Rapid prototyping (RP), has been used as an alternate means to reduce the product development cycle times, by allowing the realization of physical prototypes directly from a **CAD model** without the requirement of any tooling. Soft prototyping is the term used to refer to digital mock ups created by **CAD tools**. Recent advancements in virtual reality technology are enhancing the scope of soft prototyping.

Work is going on to develop and integrate hardware and software tools into a desktop facility that would enable the rapid development of new products, facilitate innovation in product design and allow the concurrent development of parts and the commands to realize it. The broad objectives of the above project **RAPID (Rapid product innovation and development)** are to integrate reverse engineering with rapid prototyping and to facilitate an integrated environment for virtual and soft prototyping. More specifically, our progress in the following tasks related to the realization of **RAPID** depends on:

- New **CAD tools** that are turned to layered manufacturing processes
- Interaction in virtual reality environments

- Automatic construction of **CAD Models** from physical prototypes
- Automatic planning for layered manufacturing

Motivated by certain situation in stochastic manufacturing systems, we look into the problem of optimally selecting among different input streams of waiting jobs in order to maximize the total profit in the system. We approach the problem using a multi-class queuing model with arrival streams generated by a **Markov modulated Poisson process**. We consider two separate formulations of the optimization problem. The first one seeks to maximize the profit per unit time whereas the second one considers the maximization of the profit per finished part. In either case, we explore two separate projects bounding the utilization factor (the solutions are given by especially linear programme) while the problems per finished part relationship between the solutions of the per unit time problem make the server work at a utilization factor that is no less than that of the per finished part problem.

Scheduling is an important but inherently difficult problem in discrete manufacturing systems. A large number of techniques have been proposed over the years for a variety of scheduling problems arising in the manufacturing context. We may consider dynamic and stochastic scheduling of a manufacturing system which produce multiple types of parts with significant setup costs/times involved in switching over from one part type to another. We can model the scheduling problem of a multi-class stochastic manufacturing systems, in heavy traffic in the presence of setup costs (times). It is well established that the queue length process in heavy traffic converges weakly on switching control of these limiting diffusions. The problem is formulated as a multi-parameters stochastic control problem embedded in continuous time where the evolution of any form follows diffusion dynamics. An index strategy, which is dependent only on the position of underlying diffusions, is evolved for the bandit problem. This index can be reinterpreted in terms of the original queue length process.

12.1 AGILITY IN MANAGEMENT

The globally corporate world is witnessing a sea change in the manner business is conducted today. Timely efforts are necessary to stay in business. There are no more political boundaries to restrict growth of market size. Corporates are moving globally to source their needs and supply products to countries wherever they find advantages on costs and therefore technology is moving freely across the globe. For example, **Daimler Benz**, the quality producer of cars, never went out of Germany for manufacturing activity in the past. Now they have plants in China and in India for manufacturing their products.

With the demands mounting on customer-oriented products and an ever-changing competition posed by the concept of agile manufacturing, it would be a Herculean task for our industry houses to face the challenge unless we produce goods in a record time in response to the fast growing market. This is possible only when our industry houses stand up to the challenge. Hence, there is a necessity to practice the concept of agile manufacturing in our industry houses. While liberalization has brought in a sense of discipline in our industries, it has caused anxiety in the industry houses. Unless we stand up to these demands in international market, there is fear that we might even lose our own domestic market to the multinationals. Hence, practice of agile manufacturing has become vital and mandatory.

Role of Information Technology

Moreover, a revolutionary role has been played by developments in communication technology bringing the business community together in a global village. Shape of things to come in future are indicative of people doing business from their computer terminals–ordering design of product in one country, transferring the design to another country for prototyping, ordering parts to another country and assemble in the country of use or sell globally. This appears to be a fantasy to us but we are moving very fast towards integrating ourselves with global economy.

Indian export of auto components is growing at 25% every year. It is likely to continue to grow even more rapidly. All global auto manufacturers have their presence in India now to meet the needs of local market as well as have a base of competitive manufacturing for serving other markets. We are all aware that our exports are suffering today more due to infrastructural problems than quality requirements alone. These are all signals of changing economy and market forces.

Thanks to economic liberalization in India, currently our **GDP (Gross Domestic Profit)** growth rate is 6.5%, exports growth 20% and industrial growth 12%. It is expected that if everything goes well, India is able to become the fourth economic power in the world. The environment is conducive to growth and excellent opportunities are in front of us. Only if we could make our business enterprises globally competitive, and manage the change effectively we can encash these opportunities. If we do not become competitive, someone else will. (There is no option but to change.) Recent announcement of **Y2K** to come to India created panic waves amongst the **zip manufacturers in India**. How long we can look to the government to protect us? It is high time that the industry looks inwards and prepares itself. Industry that does not have technological advantage over global competitors, has to seriously work for international competition.

Vision of Agile Management

The genesis of competition comes from meeting customer needs. As a human being, a customer has options to choose. He would like to exercise this option, if an opportunity is given to him by the market. This has brought in a tremendous amount of change in the customer behaviour and wide variations in his likes and dislikes. On the other hand, the industry has to offer wide variety of products with competitive quality, competitive price and competitive delivery to their customers. We have a classic example of the watch industry in India. When **Titan** watches came, they offered choice to customers and today they are leaders in the country. To stay in business one has to earn customer satisfaction before earning profits for the company. This obviously requires a company to be innovative, competitive and flexible. Above all it has to earn profits to satisfy the needs of the organization, its people and all the stakeholders.

There are several views about defining what makes an organization competitive. Some believe that it is the financial management that makes a company competitive. These people believe more in investment control, currency fluctuation management and interest costs control. Some others believe that it is the labour cost that makes a company competitive. Some others say it has easy access to natural resources. All of them are partly true. The competitive advantage that we are looking for, comes from **Productivity of the Business Enterprise.** We are looking for productivity not in terms of output/input ratio alone but in terms of **Results** compared to **Resources Input Per Unit** of time of the organization.

For a highly successful business enterprise, productivity in terms of PROFIT per unit of time will determine its profitability of resources employed. The higher this ratio more competitive the enterprise would be :

$$\text{Profitability} = \frac{(\text{Selling price} - \text{Cost price})}{\text{Throughput time of one piece from start to finish}}$$

To elaborate let us take an example. There are two companies A&B making refrigerators. Selling price of both brands is Rs. 10000. Cost price of company A works out to be Rs. 9000 per unit and its flexible manufacturing setup produce refrigerators within two weeks. Cost price of company B works out to be Rs. 8400 and it is able to produce the refrigerators in four weeks time. Profitability of company A works out to be Rs. 500 per week per refrigerator compared to Rs. 400 per week per refrigerator.

In spite of higher cost of manufacturing of company A by about 10% compared with company B, company A is earning more profit per week. Due to short lead-time, company A will have much more flexibility to meet changing market needs and hence more competitive. Also during recession, company A will have lesser difficulties in managing their business. It is the vision of agile management.

It is therefore clear that to make a company competitive one must focus on shortening throughput times to their manufacturing systems, quicker setup changes to accommodate variety of parts and internal setup time to change to external setup time. Distances travelled by parts have to reduce substantially. Total time the part is on the machine has to go up to 60% from the current figure of 5% to 6%. All these efforts will reduce lead-times substantially. Very often people at top have a feeling that making an organization agile needs high investments. It is not true. With very little investment the organizations can be made agile. Heavy investments can be made if higher degree of agility is warranted and business growth is able to sustain it.

Very often we put our blame of low productivity on the labour force. Little we realize that 80% of the work has to be done by white-collar employees for improving productivity. Therefore, it is necessary that white-collar employees are engaged all the time in thinking about improving productivity rather than cost cutting exercises. With increase in productivity, costs will reduce.

Upgrading, modifying and buying new hardware alone is not the answer. The agility of the organization lies in the business processes that deal with **Technology Flow and Production Management Flow** through the organization. If the speeds of these processes do not match with that of hardware, there will be no gain. Technology flow includes speed of processes like Product Design, Process Design, Work Design, jigs and fixtures. Production management flow includes speed of processes like Production Plan, Procurement Plan, Scheduling and other IT related software in the organization. All these processes are designed by people for use by people. Very often these systems and procedures are designed based on mistrust people have amongst themselves. These processes have to be simplified and more result oriented. A large amount of record is kept to spot who made the mistake? With shortening of cycle time, we will have to empower people and trust them.

Here comes the importance of flattening an organizational structure to add speed to flow of decision to the operating people. Implementation of cross-functional work teams is necessary to handle issues with common understanding and speed. Multifunctional responsibilities get added to task force leaders to achieve their objectives. Unfortunately, in our country, we have mixed up designation of a person with his emoluments and status in company and not with his work content. Better we correct this situation soon.

Any discussion of agility in an organization is incomplete without a discussion on development of agility of human mind in the organization. These are the people in an organization who make the difference. To win a race, the jockey is more crucial than the breed of horse running the race. We have to develop people who can think fast, think positive, think of progress, think of improvements, think of improving productivity all the time, and every time, think of togetherness and finally believe in strength of people rather than their weakness. Without this, no effort on business re-engineering will succeed. There are several tools and techniques that are available for this purpose but the responsibility of this change lies directly with the Chief Executive Officer himself. One cannot buy this organizational change but has to practice. One cannot increase the speed of this change. It picks up momentum depending upon the sincerity of purpose, clarity of goals and value-based operations.

Everyone is aware that it is necessary to bring cultural change in the organization. The number of training programmes being organised on this subject will probably be the largest in the world. However, the fact remains that the rate of change is very slow. In my view everyone wants others to change and not himself. If we are going to wait for others to change, the change will not take place. We have to start with ourselves and start with a programme of self-change. This will sow the seeds of vision of change in the organization and the change will come quicker.

12.1.1 The Art of Managing Creative Workforce

Managing creative people is difficult. For any organization creative people are an asset. Getting the best out of them is a challenge to any manager as people do not easily get into the system. But some managers resent creative people. They do not understand them and believe they will be better off without them on the rolls. The managers are wrong, though creative people do require special treatment. It is important to get the best out of them. Managerial psychologists, **Feinberg, Tanofsky and Tarrant,** give a few tips on how to go about that:

- **(i) Gear pressure to goals:** Many creative people need the reassurance of gentle pressure on the reins. Not steady pressure but an occasional touch. The creative person often works best in fits and starts. He needs latitude and is not easy under constant scrutiny.
- **(ii) Maintain contact:** As **Dr. Donald Pelz of Michigan University** says, creative performance is highest when supervision provides an intermediate degree of independence. For some people, this is just the entry ticket, but the creative person must not be allowed to delink contact with his superior. He benefits from knowing that there is a sympathetic and interested ear.

(iii) **Make clear the status:** The creative person is eager to be evaluated. But appraisal is difficult. It is probably futile to evaluate his procedures and results may take some time. So, the person may feel let down. But the manager must show continuing interest and redefine objectives often.

(iv) **Provide time:** For many of us, it goes against the grain to see an employee sitting with his feet propped up, staring into nothingness. But this is when the brightest ideas occur. This is not to suggest that creative persons wish to or should work in seclusion; sometimes scientists, meeting after individual studies, stimulate each other and accomplish more through synergistic effect than they would alone.

(v) **Offer job security:** A creative person loves challenges. But he responds best in security. His self-confidence, often vulnerable, is his protective shield for the sensitive mental dynamics. **Edwin Tolman** said no creative individual can function well when he is too frightened, insecure or emotional.

(vi) **Tolerate failure:** A creative atmosphere needs that a man be able to present radical, even unworkable ideas, without being harshly judged or criticized. When we ask the person to experiment, innovate and create, we must give a greater margin for error than in more predictable occupations. The manager must let this be known by word, deed and attitude.

(vii) **Respect external stimulation:** The creative person is often more oriented towards his profession than to the organization. This may clash with some fixed ideas about loyalty. Non-company activities with non-company people are important refreshers and he needs ego-building identification with an elite that transcends business. Some organizations discourage such activities which is wrong and short-sighted. Managers must accept that gifted people cannot find full satisfaction within the organization.

(viii) **Provide creative climate:** Psychologists say that men do not produce ideas when they are uncomfortable. Creative people grow uncomfortable and unhappy in a 'rule' or 'by the book' environment and pointless bureaucracy. It makes sense to relieve creative persons of routine chores–the filling out of certain kinds of reports, for example.

(ix) **Don't demand total creativity:** We can't expect a creative man to come out with ideas every moment. In fact, a hundred per cent creative man with no other attributes is not likely to be an asset. The man with something else to offer will work out better in a business organization, even if he is less intelligent than the pure thinker. **David Ogilvy** has made this point well: It is usually useless to be wholly creative. In addition to being an original thinker, you must be able to sell what you create.

(x) **Evaluate the product quickly:** Creative persons are more impatient than corporate executives. They like and need to have their contributions evaluated quickly and get fast feedback and recognition. Of course, premature disclosure on an idea may be unfortunate and sometimes dangerous. The manager must, therefore, take the necessary time to contemplate the creative man's suggestion and make it known.

Do not deny innocent and harmless foibles (old habits). It costs little to overlook oddities and it pays off, because they might just be essential to creative functioning.

12.1.2 Agile Manufacturing Needs Value-Oriented Management

Per Groholt of Norway who has written the foreword to the brilliant book **"Management by Values–Towards Cultural Congruence"** by Prof. S.K. Chakraborty of IIM, Calcutta, says, **"Prof Chakraborty** exposes the tragedy of India losing contact with her roots by importing western ideas of management and leadership. It is ironic that as the west begins to discover what the east has always known, the east is in danger of changing and forgetting"!

Prof. Chakraborty distinguishes the values from the "deep" structure rather than that of only the surface structure. Based on research on Vedantic, Buddhist, and yogic philosophy as well as Puranic literature, the values rooted in the deep structure of Indian culture and society identified by him are:

(i) The individual must be respected not because of his or her individuality but because of the transcendent, the divine enshrined in him or her, whether good or bad, older or younger, rich or poor.

(ii) Cooperation and trust–because the divine inner being of all individuals is a unity, deception or deprivation of others is deception or deprivation or deprivation of oneself.

(iii) Jealousy is harmful for mental health.

(iv) *Chitta-Shudhi* or purification of the mind with the noble thoughts of compassion, friendliness, humility, gratitude, etc. These *bhavanas* lead to a refined and accurate perception of human relationships, contributing to sounder decisions.

(v) Top-quality product/service which is primarily a function of the quality of mind or consciousness of the doer and only secondary of the techniques adopted.

(vi) Work is worship–Because the best way to approach the divine through secular life is to offer each piece of work–mentally–in as complete, perfect, humble and pure a form and spirit as one offers a flower or a fruit or a sweet to Him. This can stimulate work-ethnic in the healthiest way.

(vii) Containment of greed whether of tangibles, such as money or intangibles such as praise, because it causes stress and robs the individual of wisdom.

(viii) Ethico-moral soundness–because every action of *karma* is a cause for subsequent effect, wholesome as well as unwholesome and also because ethico moral soundness gives peace of mind and promotes mental health.

(ix) Self-discipline and self-restraint–Because they conserve energy, strengthen will-power, create trust and confer dignity.

(x) Customer's satisfaction–Because he is the divine, come upon us in human garb.

(xi) Creativity–Because human creativity is an integral component and extension of cosmic creativity; but this link has to be experienced and cultivated through mind-stilling (concentrate).

(xii) The inspiration to give–As opposed to the motivation to need, grab, etc. because giving is more fulfilling, it adds more meaning to work and life. Besides, giving with humility is more dignified than petty needing.

(xiii) Renunciation and detachment–not from duties and responsibilities but from results/rewards and egoistic demands in the workplace and from the lower, unregenerate ego and its vanities.

Organizational values essentially stem from individual values, rather than the other way around. As **Prof. Chakraborty** has analyzed many of the declared organizational values often come to nought because individual values–revealed through the quality of interpersonal, interdepartmental, inter-divisional relationships–are in a mess. The individual is thereby implicitly invited to avoid looking within, throwing the burden of such introspection on an abstraction called organization.

In fact, during the **National Conference** on **'Leadership in the Shastras'** held at Kolkata, many interesting thoughts came up on how valuable insights for Indian managers can come from the Vedas and the epics. The leadership as propounded in the Gita and the Gita psychology of ethics is based on the distinction of the **three gunas–sattva, rajas and tamas** and their interaction. As it was aptly put, **"Shastra is the system's approach in human quality improvement"**. Thus, the need of the hour for Indian managers is to make an earnest study of what our own heritage offers.

12.1.3 Agile Manufacturing means an Effective Manufacturing

In technologically oriented industries, the management discipline and philosophy has undergone a metamorphosis with the stress on motivation, leardership and organizational development. Management is no longer a simple function–specially management of manufacturing in which skills, knowledge areas and intellectual disciplines are critical factors.

The required skills can be grouped into the following three categories:

(i) **Human relations skill** i.e., the ability to work effectively in a group and to develop cooperative effort within a team and inspire confidence.

(ii) **Technical skill** representing specialized knowledge and the ability to use the techniques of the discipline. This involves working with "things".

(iii) **Conceptual skill** involving the ability to see the enterprise as a whole; the realization of how the functions of business are dependent upon each other, how change in one results change in others and the relationship of a business enterprise to industry as a whole and to the community.

Similarly, knowledge can be grouped into the following categories:

(i) Knowledge of business function: This covers items such as engineering, marketing, manufacturing and finance

(ii) Knowledge of products including their reliability and quality characteristics along with the design aspects, end purpose and performance standards required

(iii) Managerial knowledge comprising intelligence and logical criticalness besides awareness and understanding of major managerial functions.

(iv) Intellectual disciplines are basically academic, such as engineering, political science, economics and psychology. These are required for making quick decisions and correct appreciation of problems.

The raising of manufacturing effectiveness by managerial and technological means can be achieved only by the recognition of the need for improvement of skills, particularly in manufacturing management. There has been a growing awareness of training needs in the past decade and steps have been taken to meet them by government and industry. Still much remains to be done in this direction.

In the field of managerial training, executive development programmes are conducted by various organizations as well as by some of the companies themselves. But the impact of such programmes is still difficult to assess. Also, the training imparted is not oriented to Indian conditions.

In manufacturing management, the training has to be so patterned as to give over-riding importance to the need to combine the discipline and creative power of engineering science, with a deep knowledge of the human sciences and the importance of serving human values.

The time has come for education to be oriented more towards the world of work towards industry. At the university stage, the course structure should give adequate emphasis to manufacturing management in addition to its engineering core subjects. Integrated industrial training should be normal features of such courses. In order to give young people and their teachers a real feel for industry, experienced engineers should go to colleges and universities, explain the role of industry in society and cooperate with educationists in developing better ideas of work experience for potential managers and engineers. The key to education, particularly in manufacturing management, should be the word "Entrepreneur". A partial synonym is "initiative". The fact that we can only survive through innovation and productivity must be driven home.

The best engineers in manufacturing industries are those who have had good shop training to temper their technical knowledge, their judgement and their relationship with non-managerial personnel.

For the future development of the engineering or manufacturing management profession, a three-stage model along the following lines is worth considering:

- The first stage is the cognitive stage where a minimum amount of knowledge is acquired before the neophyte (novice) is exposed to the real world of industry.
- The second stage is one of planned experience under supervision, during which the knowledge and concepts gathered in stage one are applied and tempered by the realities of industry and the influence of other disciplines.
- The third stage is a monitored but unsupervised period during which the performance of the young professional is measured against previously determined and agreed objectives.

The manufacturing scene is rapidly and dramatically changing on both management and technological fronts. The need for an integrated approach to production which considers manufacture as a complete system and not a combination of separate activities is becoming explicit. In this context, it is necessary for an engineer to go through a series of special courses, both to get updated and also to widen his range of competence to fulfil his new tasks.

Concurrently with the setting up of an education system on the lines suggested, it is necessary to upgrade the quality of intake into production and manufacture. The need to encourage bright students into the line is equally great, but only when the rewards and the status accorded to the engineer are high. Today, the financial rewards are generally not attractive and advancement is considerably restricted when compared to many alternatives open to young men.

12.1.4 Management Strategies for Managing Competitiveness

The **National Management Convention**, organized by the **AIMA (All India Management Association),** was held for the first time since its inception in South India, at Bangalore on April 9-10, 1993. The proceedings were hi-tech, the most prominent being teleconferencing between Washington D.C. and Bangalore beamed live to the audience.

While metaphors were drawn to the bullfight and the battlefield in terms of managing competition, many speakers at the convention emphasized the meeting of customer needs and said that the purpose of competition should be to give a better deal to the customer. Towards this end, it was argued that there could be cooperation alongside competition.

With regard to technology, **S.N. Mishra, President of Bharat Fritz Werner,** called for **"Concurrent Engineering"** which is a part and parcel of agile manufacturing instead of sequential engineering, for rapid product development. For reducing costs, he cited that the influence of design is as much as 70 per cent and once the design is frozen, there is little room for cost reduction.

During the convention, five routes of competitive approach were suggested through strategies relating to economic, technology, marketing, finance, human resource development and services. This is shown in Table 12.1.

TABLE: 12.1: Agile manufacturing/competitive success factors

Competitive edge via	*Factors*
Economic Growth	Total factor productivity (TFD) (including infrastructural productivity)
Technology	Concurrent engineering, Influence of design, Information technology at the shop floor and the customers
Marketing	Product features, Packaging, Trade marketing, Catering to customers, Love of diversity, Strategic alliances
HRD Organization	Competition at the enterprise level, collaboration at industry level and negotiations at national level
Service sector	Quality and Reliability, Service after marketing, etc,.

In manufacturing, interesting figures were given: Of the total 8,760 hours available per man per year for production, on an average only 2,550 (29 per cent) are used. If the gap could be bridged by better productivity, costs could be cut. "Get customers into the factory and get the production managers out to customers" was the prescription for better customer-orientation"

Information Technology is the pillar around which agile manufacturing is built. The use of **Information Technology (IT)** for better manufacturing and marketing was tellingly brought out by **Siddharth Sinha** in a case study of **Bajaj Auto**. This is a unique IT case:

(a) Online systems for monitoring shop-floor operations were introduced.
(b) Expert systems were used in fault diagnosis.
(c) To counter the problem of workplace absenteeism, automatic attendance reporting was introduced, by which, at the beginning of a shift, shop-floor supervisors could monitor the

people coming in and those absent. A quick manpower reallocation could then be carried out.

(d) E-mail with dealers was introduced for speedy communication.

The example of the **Gwalior Fair** during one of the Indian Science Congress was cited, where the customers could indicate their preferences of colour and features of their scooter. These were instantly transmitted to the factory and the customers were given delivery dates on the spot. Such methods helped improve sales during a period of severe recession in the industry.

Interestingly, the senior personnel in the factory took to computerization much better than the youngsters, contrary to popular belief! Though introduction of IT meant a Rs. 25 crore investment, it paid off in one-and-a-half years. This case is a convincing one for agile manufacturing.

As for marketing, **Utpal Sengupta of Brooke Bond** said there was no substitute for a superior product in gaining a competitive edge. He said packaging was however a prime factor. Instances were cited of the initiative taken by **Tata Tea** in introducing the **polypack** in the market. It became a pace setter. A strategic alliance with the world renowned **Tetley** helped introduction of round pack as well. This must also be accepted by the community/customers. For example, one of the parts production of **GMC (General Motor Corporation)**, USA packed the parts with a plastic which was not biodegradable. The customers/community refused to accept them and buy. So **GMC** called back all packets spread over and packed with biodegradable plastic.

Apart from these, genuine cost economy from the customer's point of view must be looked into, and the popular case of **Nirma v Surf** was quoted in this context. **"Fracturing of mass markets"** was the theme for the presentation of **Ms. Roda Mehta, Director, Ogilvy & Mather.** Citing the instances of quartz watches, ball bearings and even coffee, she sought to prove that today, branding products and seizing market share would no longer guarantee loyal customers. "Others" have become a major market player. This is the result of consumers 'love of diversity', and hence the only way to gain a competitive edge is "getting into the customer's mind" and catering to his or her preferences.

Trade as Consumer

Trade too has to be treated as consumers and so the need for trade marketing was emphasized by **Kamal Ramnath, Executive Vice-President, ITC**. Distribution was a major area in effective marketing in the seventies and eighties and this is now giving place to trade marketing. This also involves restructuring of the sales force.

With regard to finance, the need for certain basic steps at the macro level, allowing firms to compete on level terms, was emphasized by **Shiv Nanda, Director Finance, Madura Coats**.

Competitive supply of funds, both on short and long term, would be needed for **SMEs (Small & Medium Enterprises)** to perform effectively. Benchmarking to provide a structured financial approach at the organizational level was stressed by him. For the public sector, it was a tale of woes and the utter lack of autonomy was highlighted by **Jagannathan** made a spirited defence of the public sector and said that in spite of all constraints, innovations are still possible as has done by **NTPC (National Thermal Power Corporation)**, in terms of its strategic alliances.

One of the presentations at the convention was by **Prof. Pradip Khandwalla, Director, IIM-Ahmedabad**. He differentiated the strategies at the national, industry and enterprise levels. At the national level, strategic alliances have to be thought of, using the comparative advantage of nations and their political relations. At the industry level, a collective strategy for R&D, which cannot be afforded by individual units, was advocated by him through industry associations.

At the enterprise level, competitive excellence has to be built up through building up a competitive organization with quick flexible/agile response and efficient teamwork. The other HRD strategies advocated were through TQM by **Anil Sachdev of Eicher** and through giving choice to employees by **Saurab Mishra of ITC (Indian Tobacco Company).**

The service sector, which has to provide a supportive base for industry to be competitive had three interesting presentations: the first by **Udayan Bose, chairman, Credit Capital Finance Corp.,** who foresees major changes in the Indian financial sector over the next two years; the second by **Ms. Camilla Punjabi, vice-president, Taj Group of Hotels,** quoting the caption "Trade follows tourism" and illustrating with the fact that the road to Gujarat in trade via Rajasthan which is known for tourism! The third presentation relating to software, was by **Mike Shah of DEL (Digital Equipment of India Ltd)** who gave an instant **"SWOT" (Strength, Weakness, Opportunity & Threat) analysis**.

According to **Mr. Shah,** the strength is that we have started and have the attention of the international customer. The weakness is that, as usual, we are not able to put our actions together. The opportunity lies in the world now refocusing on IT and the threat is that a lot of other countries, including Russia, are looking at these opportunities too.

12.1.5 Challenges in Management at Different Levels (Ref: An article on "Transitional Challenges to Mid-Level Positions" by Jayesh Chakravarthi)

Workforce Organs

Bottom Level Position	Mid-Level Position	Top Level Position
(Carry-out Institution)	(Transitional Challanges)	(Conceive Strategies)

This is comparable to a natural diamond. A diamond is nothing but a piece of carbon, which did well under pressure. Bottom-level positions are like the amorphous carbon we find in nature. As they successfully transitioned to mid-level positions, they take the form of uncut diamonds, hard but the rough edges and a few burrs. Finally, as they move into the top level positions, they are like polished diamonds, valuable and often a master's piece.

(i) Key Traits of Successful Mid-Level Professionals

Often mid-level professionals tend to be unidimensional in their approach to work. They pour themselves into executing tasks unmindful of the need to reflect on their actions, learn from multiple sources, follow processes, balance work and life and be sensitive to people around them. The problem is not one of endowment. It is about the ability to toggle among different mindsets. Bottom level position carry out institution, mid-level professionals to be successful must possess in their armoury a brew consisting of reflection, action, experience, wisdom and values. Key to success is not mere possession but the ability to judiciously mix and match these attributes to suit differing business needs, but the areas, a professional should focus on, to be effective at mid level positions.

(ii) Comfort with Change

If mid-level position, as discussed earlier is a hub, then there are many roads from it to the destination. The professional must be prepared to take any of them based on the organization's needs. In real life, the roads represent change moving to a new geography or a role, to re-skill oneself, making sacrifices in terms of reduced financial gains; keeping in mind the end goal. Managing change is about the ability to destroy one's comfort zone and re-invent oneself periodically.

(iii) Clarity of Direction

This is about the all-important difference between effectiveness and efficiency. As **Edward Deming** said, "It is not enough to do your best; you must know what to do, and then do your best".

Most mid-level professions are action oriented. They love the hustle and bustle of transactions. Many times, they miss the wood for the trees and end up not being effective in spite of putting in large effort. Successful mid-level professions develop what **Steven Covey,** calls **"helicopter vision"**. Ever so often they are able to rise above the din, survey the landscape and set themselves a direction before plunging back into execution.

(iv) Anticipative Skills

Mid-level position workforces must possess the vision, futurology, extrapolation to anticipate happenings. Sustained agility and speed is possible only through anticipation, not merely through blinding speed and brute force.

(v) Flexible Mindset

A mid-level professional must be careful to develop multiple mindsets. Everything a professional does is sandwiched between action on the ground and reflection in the abstract. Action without reflection is thoughtless; reflection without action is passive. Reflection and action needs to be collaborative, i.e., have the buy-in of the people in the organization.

Flexibility includes the ability to shrug-off failures and move forward. Professionals who step into mid-level positions often are flush with their early success.

A new mid-level professional is akin to a boy who has entered the jungle and shot his first rabbit and thinks he is the master, because he is yet to encounter the lion. Developing humility and the ability to weather failures and disappointments are key.

(vi) Respect for People

Mid-level positions open the professionals to managing people for the first time. Eager to prove their capability they tend to veer towards being abrasive (rough) rather than gentle, quick rather than sensitive and sometimes officious (assertive) rather than friendly. Mid-level professionals must learn the art of subtlety in the aspect of managing people.

(vii) Personal Values

In the context of mid-level positions, values serve two important purposes – one, of assisting the professional in his or her decision-making process and two, ensuring that the professional's actions are consistent and predictable.

Most mid-level positions are operational, involving numerous transactions and decisions. A professional with well-grounded values will find it easy to decide "what to" and "what not to", in spite of differing situations and environmental pressures, fostering transparency and trust.

12.1.6 Managing Competitive Edge for Survival and Prosperity using Agile Manufacturing

The steps for survival to maintain competitive edge would be:

(i) To develop organization and systems that are capable of responding to the customers changing needs and economic environment with desired speed and agility

With universal access to resources and given the volatile nature of business environment, industry has to continuously adjust the speed, direction and strategies of business. Since all types of capital would flow only to those industries that are sensitive to changing customers' needs and agile enough to adopt to changing market conditions, industries would need to develop organizational structures and systems that can swiftly respond to challenges. Therefore, fixed organizational structures need a transformation to agile one that have self-renewing capabilities.

(ii) To invest adequately in R&D and technology so as to offer desired product to the customers

Development of differentiated products and services, as also time-effective and cost-effective process requires continuous R&D effort. Industry would, therefore, need to invest adequately in R&D and find innovative methods to fund this requirement. Acquisitions, mergers and alliances are often motivated by considerations of generating resources for developing competitive edge so as to maintain leadership in the given business segment.

(iii) To develop product and market strategies that meet the global requirement of customers

Marketing products globally would require development of products that can be adapted/adjusted to meet somewhat varying local needs. For this, the product design strategy has to be formulated in a broader perspective of global market requirements.

(iv) To exploit the opportunities offered by the information technology and country's home-based advantage for having competitive edge

Applications of IT would facilitate management and access to distribution, sale and supply channels. Reaching to a larger base of customers and reduction in business transaction costs are other advantages that Indian industry can gain by formulating and implementing suitable IT strategies in their business.

(v) To be vigilant about costs of products and services

Cost control would be a primary concern of every industry. Industries having global strategy for acquiring capital would be able to reduce their costs on capital account. Global operation of business could provide cost advantage due to economies of scale. Every business process has to be critically examined to enable business process re-engineering to save on costs and time. Benchmarking would be another tool to keep costs low in international competition.

(vi) To ensure international quality standards

On quality front, the emphasis has to shift from the concept of quality control to quality production, as quality cannot be controlled but can be produced.

(vii) To create networks and alliances

Networking is going to be crucial to be distinct from others. Industry is already benefiting from networking of people through formation of temporary coalitions for specialized tasks and functions and from information networks. Networking and alliances can provide the industry distinctive advantages of pooling best talent available to enable them jointly serve the customized market in most cost-effective manner.

(viii) To align HRD with business

To remain competitive, Indian industries need to refocus their HRD strategies for creativity, innovation and channelling knowledge explosion. Creativity will be the key factor for the success of any industry. Wild imaginations and unorthodox approaches also need to be tried for getting imaginative and innovative solutions. It is human resources that ultimately manage the business and competition. For encouraging creativity, HRD strategy has to align with individuals so that they also acquire a tag of being distinct.

(ix) To follow business ethics and code of conduct

Integration with global economy demands that industry should adopt business ethics and code of conduct of international standards. Prevailing level of corruption is coming in the way of greater foreign investment. We, therefore, need to go back to value systems of our ancient heritage.

(x) To manage media to industry advantage

Media management for creating a positive and distinctive image and developing close relationships with customers would be another demand on Indian industry.

In conclusion, managing industry in this new millennium would require the capability of travelling in darkness with three pillars, namely, customers, competition and change, being the guiding factors in this journey of survival.

12.1.7 Change in the Organization for Agile Management

One of the problems faced by modern organizations is the accelerated pace of change. They have to learn to manage this change rather than to react to it. Thus, there is an increasing need for **OD (Organization Development)** as a way of tackling change.

The field of organization development is still growing and developing. **OD,** as an emerging discipline, has come through diverse approaches which include laboratory training, survey research feedback and socio-technical system. The blend of different value systems has historically caused **OD** to have a mixed blend of theory and practice.

There have been numerous changes in the field of **OD** in the past decade. New technologies have been developed and new research has been conducted. **OD** is being used in diverse areas. It is becoming more international in scope. There is increasing concern about structure as well as process. Power and politics are seriously being studied and dealt with. Some of these trends are listed out:

Trends in Organization Development (OD)

- From internal focus to focus on the environment.
- From focus on process to focus on structure and process.
- From focus on management to focus to all levels (involving workers at all levels)
- Shift from narrow to broader knowledge base.
- Shift from specialist to generalist or multi-skilled personnel.
- From consultant to line manager–as change agent.
- From single interventions to multiple interventions.
- From truth and love to power and politics.
- From participation to contingency approach.
- From single country to multinationals.

- **Focus on Environment**

Traditionally, **OD** practitioners have focused primarily on the internal workings of the organization, in shaping its mission and goals and on organizational structure and design. Thus, they need to become more aware of and more concerned about, strategic and tactical planning. As change becomes more rapid and the environment more turbulent and as inflation increases, organizations must become more proactive rather than reactive to the environment.

- **Focus on Structure and Process**

OD has focused on process through such approaches as laboratory training and survey feedback. Little attention was paid to the structure of the organization or its work. It is clear that the emphasis on process, without corresponding attention to organization design and structure, is of limited value. Thus, as a corollary to the preceding trend, the **OD** practitioner needs to work with the client system to help evolve an organizational structure and/or structure of work that helps the organization achieve its purpose and humanize the members of the organization. Two relatively new but important approaches to organization structure are contingency theory and matrix organization. In both, the emphasis is on lateral rather than vertical relationship and on better ways to enable managers to perform essential coordinating functions. There is a clear need for **OD** practitioners to work with their client system in agile manufacturing to determine the conditions under which a matrix may or may not work before such an intervention is suggested. Diagnosis of the need for appropriate structure should come long before the attempt to prop up a matrix structure which may be inappropriate.

- **Focus at all levels (Involving Workers at all Levels)**

Traditional **OD** focused primarily on the target groups–management levels of the organization managers through such processes as laboratory training, team building and process consultations. In the prescriptive days of job enrichment, the workers were not involved in the redesign of the work itself. Now, through such approaches as **QWL** (Quality Working Levels) and socio-technical systems, there is a greater emphasis on worker involvement at all levels, including working directly with unions. Workers are more directly involved with the design of their own work, as a principle of agile manufacturing.

- **Shift to Broader Knowledge-base**

From a systems point of view, a change in one subsystem affects other subsystems. Key organization processes are affected by such factors as the external environment, technology, formal organizational arrangements and the social system. The field of **OD** has concentrated primarily upon separate and disparate (different/distinct) techniques. It has been inward looking and has paid little attention to other behavioural sciences, let alone other management areas such as marketing, accounting, production, finance and personnel.

- **Shift to Team Approach**

Traditionally, the **OD** consultant has worked alone or with other internal or external practitioners. Clearly, no one can understand or comprehend the modern complex organization and/or find and use the proper leverage points. Thus, the **OD** practitioner is being stressed, to work with a variety of other specialists including compensation, labour relations, strategic planners and industrial engineers. Thus, there is a corresponding need to bring a number of resources to bear on organizational and human problems.

Indeed, a more strategic approach is to work with and train expert professionals, especially those in staff jobs, so that they learn how more effectively to make their expertise available to the organization. An engineer who has knowledge of **OD** may be more influential in new plant design or new product introduction.

As a corollary to the comments made above, there is a much clearer recognition that the real change agent is the line manager. Action research is a process of struggling jointly with the customer to develop better approaches to a more adaptive organization, but the managers (and work group) are the real agents of change. The consultant can only be a facilitator.

In the beginning, traditional **OD** frequently used a single intervention as the primary mode of change. For some, sensitivity training was the answer. For others, **MBO (Managing Business Opportunity)** or job enrichment was the preferred mode of intervention. **OD** practitioners tend to do what they know best and different practitioners still have their favourite interventions. The trend shows a shift to multiple interventions.

The early history of **OD** emphasized the necessity for all organizations to have an open, caring, trusting climate, using the T-group as the primary intervention mode. As **OD** is being used in a wide variety of organizations, including the State and local governments, and hospitals and it has become more involved with unions, particularly through the **QWL** movement, there has been a growing realization that life itself is not, nor can it ever be, all truth and love. As a result, **OD** practitioners are facing more squarely the issues of power, conflict and politics.

A decade ago, books and journals announced the death of bureaucracy and the coming of open democratic organizations. All managers would be participative in all situations. The current trend suggests a situational or contingency approach in which decision-making should be more participative in some circumstances and less participative in others. Contingency approaches to organization design require contingency management. There is also an increasing evidence that the better manager is one who uses an authentic rather than a contrived style of management and that managers and others cannot be expected to change their personality or style of interacting with others.

- **Multinational Approach**

Relatively little **OD** work has been done on a multinational basis. Nevertheless, the continued growth of multinational corporations generates new managerial and organizational problems and dilemmas. In general, there are problems of differing cultural and environments, including laws and regulations, problems of organization design in a different cultural milieu, corporate bureaucracy versus local heterogeneity, career problems and planning including the strategic transfer of managers and increased need for coalition organization and management skills.

There has been only a little help in handling such problems. Thorough examination of the culture is necessary to determine proper plant and organization design, selection, recruitment and development of workers.

For illustration, the **Scandinavian** countries are involved in industrial shop-floor democracy, autonomous work groups and socio-technical systems. What differences are necessary when an **American organization** does business in **Scandinavia** or vice versa? In **Japanese organizations**, it is not uncommon for the chief executive to spend some time each day working at the shop-floor level. What might be the effect on culture and norms of such activities in a different environment?

To conclude, the field and practice of **OD** will continue to expand and become more complex, involved more closely with other disciplines, drawing more heavily upon them and requiring more variables. In a world that is rapidly changing, organizations need to become more adaptable and to better learn to manage change. From a system point of view, change is enormously complex and can come from inside or outside the boundaries of the system. A major key to managing change is proper diagnosis of problems and situations, keeping in mind that the performance of the whole is not the sum of the individual parts, but is a consequence of the relationship of the performance between the parts. Thus, problems cannot be solved separately, since they are interdependent. Experience is not really the best teacher–experimentation is. The line manager is fundamentally the agent of change. The consultant's task will be to work with the manager to experiment and try out new approaches.

In all interventions the focus should be on the job and the tasks to the extent possible. Setting of goals and priorities and determining obstacles to attaining those goals should have a higher priority than interpersonal feelings as such.

12.1.8 Response Management in Organizations

Ordway Tead defined an organization as a system of collaborative creation. Viewed as a system of cooperative activities of more than one person for the accomplishment of a common goal, an organization becomes, in fact, a vehicle for achieving social objectives. It functions in an environment characterized by external conditions and influences that affect its life and development, growth and prospects. The closer environment of an organization consists of customers, suppliers and labour. The broader environment, on the other hand, consists of competitors and trends affecting the system of organization and its working environment.

Two-way Process

The total environment of the organization provides enough agility for withstanding changes. Strategy in this reckoning is a configuration of an organization with its environment and the total formation, resource allocation and marshalling of other means for producing this configuration. This means that not only the organization reacts to change, responding to the organizational process but it also initiates changes that would contribute to its long-term well-being.

This two-way process characterizes industrial and business organizations engaged in a constant process not only of adoption to changes taking place in the external world but also causing changes that would suit its interests. Since a business organization is a social entity, at a point in time and space, the interest of the organization, in fact, merges with that of society at large-producing economic goods and services, creating surplus, creating employment and utilizing resources made available by society.

In the history of business movements, environmental change occurs, which exceeds the agility of an organization or the rate of conflict varies with the direction of adaptation. This process is called discontinuity. Where the organization is a continuing entity, more responses by members of the organization are necessary for achieving goals or strategy. Response itself is brought about by a number of sub-responses that collectively adjust the system of the organization and its environment until it again functions so that it can satisfactorily meet the goals.

Response to discontinuity, therefore, underlines actions concerning uncertainty. Deficiencies in structure, systems, intelligence and processes become noticeable in such a phase of an organization. Discontinuity jerks and jolts the processes of adaptation to change underlining the different inadequacies in satisfying the objectives of the organization.

The response process in the organization is essentially three-pronged, each having different sub-responses:

(i) Corporate response underlining adjustment of the number of strategically independent businesses in a diversified company.
(ii) Administrative response involving introduction of revised structure or new systems or processes in business
(iii) Operating response stressing amendment of the operations of an organization.

The response process itself contains six features, namely, Structural change, Goal definition, Constituent sub-responses, Source and nature of required expertise, Personnel and their interrelationships and General atmosphere characterizing a response of each type. The practical components of these six characteristic features noticed in case of three types of responses, Corporate, Administrative and Operating, are given in Table 12.2.

TABLE: 12.2: Corporate, administrative and operating responses

Components	*Corporate Response*	*Administrative Response*	*Operating Response*
Structural change	An added structure independent of the original structure	A revised structure	Expenditure around existing structure

Contd.

Goal definition	Clear ultimate goal will be expressed	Unclear goals will exist until ambiguities are resolved, then general principles will be established	Goals apparent perhaps day-to-day survival
Sub-responses	Sub-responses will follow from goals but cannot be pre-determined in detail	Sub-responses will be translated from goals in an uncertain and evolving manner	Sub-responses often will be mostly less than obvious, and extreme actions sometimes will be required
Expertise	Management expertise will have to be obtained by specific means (a project team, for example)	Management expertise will be found by revising behaviour of present incumbents	Management expertise may be challenged but will almost always be adequate
Personnel	Special or outside personnel probably will be required	Mainly the same managers with different relationships; outsiders possibly added	A set from the same personnel, perhaps supplemented with others
Atmosphere	Clear mission, but needing learning and experiment	Introspection and re-education	Clear tasks, ingenious solutions, and expedient high pressure working conditions

In strategy formulation, environmental scanning is vital so that the external stimuli are spotted, defined and delimited for generating necessary response processes. Organizational sensitivity is central to such activities. It is, however, not necessary that such response processes are worked out only for satisfying the goals of organizational members. In a corporate setup, it is agreed that the body corporate has a personality of its own and has interests discrete from the entire body of shareholders and other members of the public–internal and external–which the unit is to serve.

From this point of view, it is essential that the strategic management process is sensitized for ensuring perpetual existence of the body corporate which does not die, but which can be killed. The response process has to take into view not only the members of a company but also the company itself. This may imply responding to discontinuity by way of disinvestment, emphasis on status quo ante and expansion or diversification. **Ansoff's** idea of environment servicing organizations emphasizes that even for the so called non-profit organizations, strategic management for ensuring long-term prospects is as essential as it is for commercial organizations. It is held out in major part of management literature that organizations develop necessary strategies for profit.

Non-profit Organizations

In the context of environment servicing organizations, such an approach would naturally call for modification of approaches. While profit in the truly commercial sense may not be given the upper hand, generation of surplus–at least that much considered adequate for ensuring perpetual existence–must not necessarily be worked for. This means that even non-profit organizations vote for strategic alternatives that would ensure, first of all, no profit no loss and then, profit or, for that matter, surplus which would ensure their long-term sustenance. It is, in fact, implied that for ensuring perpetual existence, all bodies corporate must necessarily be self-sustaining. The discontinuity aspect has thus to be understood in this specific context.

Scenario Planning

Whether it is a profit or non-profit organization, for development of the right responses an organization has to dabble (take part) in what is called scenario planning which would require assessment of the vogue (current) phenomena in the environment, anticipation of the changes in the phenomena in the near, not-so-near and distant. Wherever indications are available, whether firm or faint, an organization has to show alacrity (eagerness) to spot changes and to define, delimit and delineate its action programme for meeting the challenge.

Where causation (effect) of a change is contemplated, such scenario planning would help charting out the courses of events in the external environment areas where changes are visible and the commitment of resources required for initiating such changes. The business world is perpetually engaged in a game of chess. Not only there are competitors in the same line of activity but there are substitutes for similar end-uses and change in the existing end-uses may mean massive effort. Drawing of the scenario, the multitude of problems visiting the environment and the manner of dealing with them would also require a clear perception of the problem as such and the manner in which these problems could be dealt with for benefit to the organization.

Scenario planning would require putting on the canvas both problems and prospects and the methods of dealing with both. Stopping with the problems many mean prospects may go amiss. Dealing with prospects only may imply that the problems would provide a counterpoise to the prospects sought. In the circumstances, the hurdles in the way should be identified before they are sought to be removed so that prospects materialize on the lines anticipated. This is no easy task but on this would depend whether an organization succeeds in its venture and whether the objectives set would be realized or otherwise.

The response process as indicated in Table 12.2 of corporate, administrative and operating resources would all be relevant in such scenario planning. The timing of such response is of crucial significance and the ultimate success of an organization would depend on the relative costs and benefits of the measures taken or contemplated. It is essential from this point of view that the organization remains alert all the time and does not rest on its oars. It has been said that most organizations fail to score on this count.

12.1.9 Time: A Valuable Tool for Competitiveness in Management

Matsushita now produces a washing machine in just two hours as against 360 hours a few years back. **AT&T** takes only one year now to design a new telephone compared to two years earlier. When **Yamaha** challenged **Honda's** supremacy in the motorcycle market in the early eighties, Honda's reaction was with the introduction of 113 new models thereby effectively tuning over its entire product line twice in just 18 months.

Benetton, the largest ready-made garment manufacturer in the world with extensive sales outlets around the globe takes hardly 15 days to receive the order and deliver a fashion garment of one's choice in the midst of a busy season. Closer home, there are new niche markets with one hour photo and two minute pizzas emerging to meet the requirements of customers who want faster service. Time is fast emerging as the scarce resource.

- **Survival of the Fittest**

Recent years have witnessed emergence of a turbulent environment all over the world for business to operate, due to both economical and political reasons. This is especially so in the Indian context with the opening up of the economy to international competition and rising consumer expectations about the quality and variety of products which would meet their specific requirements. Since customer specifications vary with individuals, there is greater pressure on manufacturers and suppliers of goods and services who lie with each other for survival and growth in the marketplace to compress time to react to the situation. Under such conditions, competitiveness depends on the ability of firms to cut their operating cycle time to the minimum. The situation is that where time is more critical or scarce, a resource is more important than often money. Some of the possibilities of achieving this end are examined here.

- **Changing Environment**

The eighties marked the beginning of major structural changes in the Indian economy with the entry of several automobiles, electronics and computer firms in the light of the sudden changes in government policies. The entry of automobile firms with strong **Japanese collaborations** forced the existing manufacturers to wake from the slumber which they were used to. **Bajaj Auto** was shocked to realize that its waiting line for scooters had started disappearing suddenly.

The mid-eighties saw Bajaj bringing about major strategic changes. The result, Bajaj gave place to new humble, market oriented ever smiling Bajaj. Now not only that the factory is highly automated, but its dealers can place orders through centrally connected computer terminals and be sure of getting delivery of vehicles of their choice in a few days.

Consumer electronics also passed through the early stages of growth and reached a plateau where product features deserved attention. Manufacturers started luring consumers to their brands by offering different features to take care of individual consumer needs. This scenario of increasing competition and the need to cut costs and at the same time provide variety in terms of product features and colour can be analyzed under differentiation and cost leadership (cost reduction).

- **Differentiation**

Basically individuals are different and if they have been tolerating the same kind of TV or car for years together, it was not because they never had individual tastes and personal preferences, but because they had no choice. The policies of the government which limited entry of new manufacturers into the field never allowed consumers to have a choice.

The government's thinking seemed to be quite blank and products were to be purely functional and nothing more. Now that competition is allowed and new firms with success strategies can penetrate into the market, customers can put their foot down and ask for what they want. The situation is fast emerging like a restaurant hall where the customer can place an order and get served according to his specifications.

It is precisely for the capability to react to individual customer requirements that Benetton, once a small garment manufacturer in a remote region near Venice in Italy became the largest garment manufacturer in the world. The company could compete with products from low cost economies such as Taiwan, Korea and China although they are located in a high wage economy.

To compete in a matured product market such as garment could not have been possible but for the capabilities the company has developed to meet individual fashion preferences within a period of 15 days of ordering at the retail shop. This means that a customer walking into a Benetton retail shop anywhere in the world can place orders for shirts of his specifications, get them manufactured in the company's factory located thousands of miles away in Italy and can get them delivered within two weeks at the retail shop.

Same is the experience of Toyota which is now able to deliver within three days a car as per the feature requirements of the customer from its factory in Osaka. Since such speedy service does not result in extra cost, there are always customers lining up in the retail outlets or showrooms of such companies.

- **Cost Leadership (Cost Reduction)**

Conventional manufacturing has shown that cost advantages are possible by increasing the volume of production. This was the basic premise on which even the JIT philosophy got evolved leading to emergence of focused factories where a factory would specialize in a single product and produce as many pieces of the same thing as possible repeatedly.

Automation and robotization were subsequent developments created based on this paradigm primarily to bring down cost per unit of output. Since JIT essentially means elimination of all kinds of wastes, unnecessary build up of raw material, work-in-process, finished goods and even operations or movements are removed gradually. Achieving 100 per cent perfection takes time, but that is the goal one aims to achieve.

The overall result is that production takes place in response to orders, thus effecting a suction from the market as against conventional manufacturing philosophy of pushing products from shop floor to market. At the micro level, the assembly line would not assemble and keep finished goods in store waiting for orders to come. Assembling is done as per customer requirements. Extending the suction argument to the machine shop, no part is machined and kept in the store waiting, instead machining is done as and when required. This approach to manufacturing achieves higher employee morale through teamwork, productivity and waste reduction also.

- **Lean and Fit**

Cutting costs is important not only to have maximum economies, but also to be as lean as possible to face competition and possible hard times. It is only a physically fit and trim person who can fight his enemies with quick reflexes. Cost cutting also results in improved earnings as the market price is not affected by it. The result is increased staying power and more resources for expansion, market investment and R&D.

In the general management paradigm, cost leadership and differentiation are two divergent situations mainly because cost leadership means higher volume and larger market, and differentiation means market segment and variety. The former situation assumes large batch size and the latter small. The latest revolution in industrial world has achieved a blend of the two resulting in the emergence of flexible factories. Firms with flexible factories can meet the demand from any point on the market spectrum which has at one end customers with minimum individual specifications and at the other end customers with maximum individual product/service specifications. This means that one can follow cost leadership to fight competition in a matured marketplace and at the same time, be successful in a premium price market segment where variety is the major success factor.

- **Criticality of Time**

In a turbulent environment where consumer preferences undergo rapid changes, and competition high, it is important to identify end-user requirements and provide the services at the earliest to be competitive in the marketplace. This essentially means that firms have to identify ways of compressing time to complete an operating cycle starting from the point of taking orders and ending with delivery of the product/service.

There are two prerequisites to compressing time.

One, each of the operations required to be completed in the making of a finished product or service should be arranged sequentially in such a way as to ensure free flow without any restrictions or obstacles. This essentially means that under suction or pull from the succeeding operations, parts or components are processed and moved on. It can be likened to flow of water under gravitational pull from below.

Two, perfect coordination will have to be maintained between each department and among operations within each department so as to minimize wastage in terms of waiting for activity to begin or to complete.

Compressing time raises a fundamental question as to what is time after all. Time is the duration of an activity. It can include a phase during which one process is over and yet other one has not begun. Compression of time therefore means compression in the duration of the activity. This is possible only if the need for each activity and the assumptions behind each activity are examined in detail.

Very often, the need for any of the activities is decided based on conventional manufacturing principles followed and practiced over the years. The assumptions behind any of these activities especially in terms of the need for each of them are normally not questioned. It is time that one started questioning the need to have any single activity involved in the manufacturing of anything and find out new methods either in terms of new technology, equipment or in terms of new processes which can compress time. For instance, one can ask why should the casting of a simple 0.5 HP motors have x number of nuts and bolts instead of x 1 or $x + 1$. Automation of several activities especially in engineering and electronic industries has resulted in considerable saving in time.

- **Questioning the Assumptions**

At another level, any activity in a business context is based on at least one, if not, several assumptions. For instance, the capacity of a plant or the productivity of labour is based on the assumptions which in turn are based on previous experience. There is nothing to prevent to change many of these assumptions or at least questioning these assumptions so as to find out where improvements could be made. Behind any innovation, one can locate an inquiring mind, one which questions previous assumptions. Changes cannot be introduced without changing many of these assumptions which are either controllable or uncontrollable in nature.

Generally, assumptions about labour productivity or availability or managerial time or internal funds availability are all considered to be controllable by the firm. A closer look at any of these would show that there are innovative ways of bringing about changes in labour productivity or even raising additional funds.

For instance, the shift from focused factories where one factory specialized in the mass production of a single product in the shortest possible time and thus cut costs of production, to flexible factories where while maintaining the advantages of focused factories, an additional feature to provide a variety of products to meet customer requirements is added. This was the result of increasing competition and the need to meet specific customer specifications.

Uncontrollable assumptions are those which are generally believed to be beyond the control of the manager. This is not entirely true mainly because some of the uncontrollable assumptions can be moved closer to controllable assumptions. For instance, the sub-contracting firms outside the control of the mother unit can be brought closer to the category of controllable assumptions by having managerial and ownership interest in them or through technological support or at times emotional attachment.

Benetton has brought its independent retailers with whom they do not have even a franchising agreement closer to a controllable stage mainly by providing them facilities to make changes in their orders in terms of price, colours and designs in the middle of season. This facility enables the retailer to increase his turnover and be loyal to Benetton as no other fashion garment firm could offer such a facility. There are also firms with lassoing (rope with a noose) facilities to influence even government regulations and policies which are supposed to be completely uncontrollable.

The basic lesson that one learns from the above discussion is not so much in terms of the tools and techniques of modern management, but how to be competitive when time is a scarce resource. One should learn the rules of the game as described in the textbooks, but the innovative and the fast growing firms should learn to break these rules and be entrepreneurial to be more competitive.

It is this process of breaking rules which results in questioning the assumptions and making changes in the product as well as processes that make an entrepreneur. Compressing cycle time and making firms to build competitive advantage on the basis of time are possible, but they need a creative and innovative mind.

12.1.10 Competition Unmatched

To stay ahead of competition, manufacturers are accelerating their operations and improving the quality of what they make and sell. For example:

(i) **General Electric (G.E.)** used to take 3 weeks after an order to deliver customized industrial circuit-breaker box. Now it takes only 3 days.

(ii) **AT & T** has reduced its time of 2 years to design a new phone to only one year.

(iii) **Motorola** has speeded up its turnout of electronic pagers from 3 weeks to 2 hrs from the date of order.

Quicker developing, making and distributing products or services being important, sometimes surprising competitive benefits. Market share grows because customers love getting their orders at the earliest. Inventories of finished goods shrink because they are not necessary to ensure quick delivery. Fast manufacturing can make and ship an order the day it is received. For this and other reasons, cost decreases. Many employees become more satisfied than delayed supplies because they are working for a more responsive and successful company, as the rapid operations required give them more agility and responsibility. The experience of **Hewlett-Packard is "Doing it fast forces you to do it right the first time"**.

The investigations reveal that speed often phases off in product development even if it means going over budget.

Little wonder that corporate speed is becoming a principal topic of study among management consultants and business school professors. It has the conceptual heft to influence virtually every operation in every company.... "Thinking in terms of time" says **Rudyard Istvan**, a Senior Vice President at **Boston Consulting Group**, "gets all.. activities working together for a common purpose".

A company can generally find thousands of ways to speed operations, but masters of the game emphasize a few major tactics. Among them are for example:

- **Start from Scratch.** Practically everyone agrees that the worst way to speed up a company is by trying to make it do things just as it does, only faster. The machinery–and certainly the workers–will simply burn out. Instead, smart executives announce a breathtakingly ambitions goal, like **Young's** target of halving break-even time. No one can get there just by stepping on the gas, although some try. "The gut reaction", says **Steven Wheelwright, a professor at the Harvard Business School in Cambridge, Massachusetts,** "is let's cut six months out. It's not, let's do the analysis and rethinking that will allow us to cut six months out."

  But that's what it takes. Example: **General Electric's** rescue of its old and stagnant circuit-breaker business. Says **William Sheeran**, a **GE** General Manager, "we had to speed up or die". **GE's** $1,000-million-a-year electrical distribution and control division in **Plainville**, Connecticut, makes, among other things, circuit-breaker boxes for commercial buildings. When threatened in the early 1980s by the market's slow growth and tough competitors like **Siemens and Westinghouse**, GE formed a team consisting of manufacturing, design, and marketing experts that focused on overhauling its manufacturing process. The goal, suitably daunting, was to cut the time between a customer's order and delivery from three weeks to three days.

  Time for radical thinking: GE was producing circuit-breaker boxes in six plants around the United States. Who needs six? The team decided that one plant would be more efficient, so in 1985 it consolidated in operations and focused on automating its factory in **Salisbury, North Carolina.**

  But the team did not want to automate operations as they were. In the old system, engineers custom-designed each box, a job that took a week. They chose from 28,000 unique parts to create the boxes....Setting up an automated system to handle that many parts would have been a nightmare. So the design team made most parts interchangeable, reducing their number to 1,275 while still leaving customers a choice of 40,000 different sizes, shapes, and configurations of boxes.

The team also devised a way to cut out the engineers, replacing them by a computer. Now when a salesman enters specifications for a circuit-breaker into a computer at GE's main office in Connecticut, the order flows to a computer at Salisbury, which automatically programme the factory machines to make circuit breaker boxes with minimum waste of material.

Impressive advances, but the team still had to conquer another source of delay-solving problems and making decisions on the factory floor. The solution was to get rid of all line supervisors and quality inspectors, reducing the organizational layers between worker and plant manager from three to one. Everything those middle managers used to handle such as –vacation scheduling, quality, work rules–became the responsibility of the 129 workers on the floor, who are divided into teams of 15 to 20....The more responsibility GE gave the workers, the faster problems got solved and decisions made.

Today the **Salisbury plant** basically runs itself at a rapid clip. On the factory wall a giant electronic sign hung eight metres off the ground flashes in red letters, letting workers know how long it is taking them to make each circuit-breaker box, how many boxes they have to make that day, and how many they have made so far. The sign lets employees pace themselves and make their own scheduling decisions. Says **Dottle Barringer, an ebullient-woman** who has worked at the plant for 12 years. "I like to be my own boss. I don't like to be told what to do. I know if I can't get it done in eight hours, I can do it in ten without getting permission for overtime. We're behind right now. No one has to tell us we have to work Saturday".

Results: The plant, which used to have a two-month backlog of orders, now works with a two-day backlog. Productivity has increased 20 per cent over the past year. Manufacturing costs have dropped 30 per cent, or $5.5 million a year, and return on investment is running at over 20 per cent. The speed of delivery has shrunk from three weeks to three days for a higher-quality product with more features. GE is gaining share in a flat market. Sheeran says, "we'd be out of business if we hadn't done it".

- **Wipe Out Approvals.** Consultants recommended that managers take a hard look at the number of times a product or service requires some sort of internal approval before reaching the customer. **Philip Thomas, President of Thomas Group, a Dallas, Texas,** consulting firm specializing in speed, says that manufacturing typically takes only five to ten per cent of the total time between an order and getting a product to market; the rest is administrative. He says, "Most people don't know how their own company works. Make a flow chart of your administrative process to show what's really happening out there. It usually shows that getting a product to market takes longer than people think".

Jack Reichert, Chief Executive officer of Brunswick, the $3,000-million-a-year sporting-goods company, decided in the mid-1980s that developing new products-outboard motors, automatic bowling scorecards, fishing reels took too long....Reichert cut the layers of management between him and the shop worker from ten to five. He says, "If you want to know how to do a job right, ask the people who are doing it; don't have some corporate bureaucracy to tell them how to do it. "Reichert also bumped up spending authority for division heads from $25,000 to $250,000. At headquarters in **Skokie**, Illinois, he gets only a weekly one-page report from each division chief. "Those reports", says Reichert, "tell me what they need". **Reichert estimates** that **Brunswick** can now develop new outboard motors, for example, 25 to 30 per cent faster....

- **Try Teams.** Like **GE** in its circuit-breaker business, nearly all the fastest companies form multi-department teams. **AT&T** is applying that approach to product development with resounding success, giving its teams deadlines and lots of authority to make the decisions required to meet them. **AT&T** used to take two years designing a new telephone. But, says **John Hanley**, an **AT&T Vice President of Product Development**, "We came to the realization that if you get to market sooner with new technology, you can charge a premium until the others follow".

AT&T began developing a new cordless phone for the home called the 4200 in early 1988. Hanley faced a rigid **AT&T bureaucracy** and was skeptical about his chances of cutting time out of the development process. Rather than trying to save ten per cent in time here and five per cent there, Hanley aimed to reduce the development cycle by 50 per cent. He says, "It made us change the way we did everything". In the past, AT&T product development worked like a relay race, with the engineering department handling a design over to manufacturing, which handed the finished product over to marketing to sell. For the 4200, **Hanley** considered setting up a...product-development lab off premises where engineers could work relatively free of bureaucracy. He soon rejected that idea, figuring that if his people couldn't develop a new product within AT&T they wouldn't achieve a significant and permanent change.

Instead, Hanley formed terms of 6 to 12, including engineers, manufacturers and marketers, with authority to make every decision on how the product would work, look, be made, and its cost. The key was to set rigid speed requirements–six weeks, say, for freezing all design specifications. Because the team did not need to send each decision up the line for approval, it could meet these strict deadlines. With this new approach AT&T cut development time for the 4200 phone from two years to just a year while lowering costs and increasing quality.

Some companies make suppliers part of the team. **Navistar, the Chicago, Illinois, truck maker,** landed a major contract in 1987 with **U-Haul**, (a company that rents trucks and trailers to the public for moving household goods), by promising to deliver a new van fast. It was a stiff challenge. To make loading easier, the truck bed had to be only 61 centimetres off the ground, compared with the standard 122 centimetres. That required an entirely new chassis design and a novel type of pneumatic suspension that lowered the truck bed ten centimetres at the flick of a switch. Says **Neil Springer, Chief of Navistar's truck-building subsidiary**, "We realized we couldn't deliver the truck fast enough in the traditional way. It had to be done in teams, with everybody, suppliers included, working simultaneously." The approach succeeded and produced a bonus. The **Dana Corporation**, a supplier that makes frames and axle, suggested a way to help lower the bed by redesigning the axle, a suggestion that would not have been heard until perhaps a year later under normal procedures. Taking advantage of it would have caused considerable delay while manufacturing.

- **Worship the Schedule.** Managers must make clear that nothing short of disaster is a valid excuse for delay. In 1988, **Motorola** designed a new electronic pager and an automated factory to build it in an unheard-of 18 months rather than the standard three years. The first step to achieving that goal was setting clear deadlines. After that came discipline. "When you're dealing on the edge of the envelope, when you're flying blind, there's always the temptation to take a little more time", says **Scott Shamlin, a Director of manufacturing at Motorola**. "But we hold to schedule as a religion". **Motorola's** automated factory in **Florida** can build and ship an electronic pager some two hours after it receives an order. The job used the U.S. pager market over Asian rivals.

- **Remember Distribution.** Even the world's fastest factory won't offer much competitive advantage if everything it produces gets snagged in the distribution chain, a lesson **Benetton,** the Italian sportswear company, has knit in its heart. Located in Ponzano, **Italy, Benetton** makes and distributes 50 million pieces of clothing worldwide each year, mostly sweaters, slacks, and dresses.

  Benetton found that the fastest way to run a distribution system is to create an electronic loop linking sales agent, factory and warehouse. If, say, a sales woman in one of **Benetton's Los Angeles, California**, shops finds that she is starting to run out of a best-selling red sweater in early October, she calls one of Benetton's 80 sales agents, who enters the order in his personal computer, which sends it to a mainframe in Italy. Because the red sweater was originally created on a computer-aided design system, the mainframe has all its measurements on hand in digital code, which can be transmitted to a knitting machine. The machine makes the sweaters, which factory workers put in a box with a bar code label containing the address of the **Los Angeles** store, and the box goes into the warehouse. That's right-one warehouse serves Benetton's 5,000 stores in 60 countries around the world. It cost $30 million, but this distribution centre, run by only eight persons, moves 2,30,000 pieces of clothing a day.

  Once the red sweaters are sitting snugly in one of 300,000 slots in the warehouse, a computer sends a robot flying. By reading the bar codes, the robot finds the right box and any other boxes being shipped to the Los Angeles store, picks them up, and loads them in to a truck. Including manufacturing time, Benetton can get the order to Los Angeles in four weeks. If the company already has red sweaters in stock, it takes one week. That's quite a performance in the notoriously slow garment industry, where hardly anyone else will even bother with reorders. And if Benetton suddenly realizes that it did not make any, say, black cardigans and purple blouses this year and they are hot, it can manufacture and ship a "flash collection" of black cardigans and purple blouses in huge quantities in a few weeks.

  Put Speed in the Culture. ...**Domino's Pizza**... has become America's second-largest pizza chain... by promising customers a $3 discount on any pie that takes longer than 30 minutes to arrive at their home.

  Chief Executive Officer Tom Monaghan... has modelled Domino's on a professional sports league, with regional divisions competing to make the best pizza, deliver it faster, and sell more than the next region. Like [football coaches], Domino's Pizza makers watch films of the fastest pizza makers in the country to learn winning techniques. For example, a delivery person, whose uniform includes track shoes, runs from delivery truck to a customer's house or apartment, taking stairs two at a time-elevators are forbidden because they take longer-carrying a pizza in his arms...Domino's franchisee in Washington, D.C., takes managers out every Monday for a ten-kilometre jog. Says Mohaghan: "Our whole business is built on speed".

The advantages of speed are undeniably impressive. But why should that be surprising? Why all the hoopla (game) among management theorists? After all, it's not exactly a new precept (new rule) that time is money.

True enough. What is new is a realization that time is an awful lot of money- far more than most executives understood. **Warns Istvan of Boston Consulting Group**: "If you come up against one of these fast corporations and you're not prepared, you're history". Those who grasp the new calculus, who appreciate the unprecedented advantages of getting new products to market sooner and orders to customers faster, may well hold the principal tool for achieving competitive pre-eminence in coming years.

12.1.11 Agile Manufacturing Challenges in an Open Regime

The new industrial policy announced by the Government and the concomitant (associated) measures in different spheres of the economy have apparently done away with some of the existing vogue, constrictive approaches. The withdrawal of restrictions promised by the government has created conditions that can easily lead to chaos. Economic planning in the country tended to protect the unfit and for all practical purposes, different official policies took the weakest links in the chain as the scales of reference–be they administered prices, official incentives, preferential allocation of inputs or other matters.

The withdrawal of permits and licenses has suddenly exposed corporate enterprises to stiff competition, not only from other enterprises within the country but also from outside. As a result, the guiding philosophy has already become 'Survival of the Fittest'.

(i) Diseconomies of Scale

A major weakness of Indian enterprises that is easily identifiable is the uneconomic scale. There are about two lakh companies at work, of which only about 25,000 are public limited companies. The Reserve Bank of India – a sample of 1,850 public limited companies comprised 70 per cent of the total paid-up capital of such companies at work. The size of paid up capital for the other public limited companies can be imagined.

There are some 1,200 government companies operating, so named by virtue of government's holding more than 51 per cent shares of such companies as defined under Section 617 of the Indian Companies Act, 1956. The total paid up capital of government companies–only 1,200 of them–was about three times that of the entire non-government corporate sector. The dismally low per unit paid-up capital of the non-government companies can be imagined again.

Even among government companies, about 240 of the Central Government controlled companies alone accounted for about 75 per cent of the total paid-up capital. The point sought to be underlined on the basis of this fact is that the diseconomy of small scale in the corporate sector has been pervasive (present).

Being inherently weak, most of these companies had so far been basking (engaging) under the protection offered by government policy. True, for certain industries, the government did prescribe minimum economic capacity but unfortunately such prescription was almost entirely in terms of production, not in terms of investment requirement. At the other end of spectrum, the government sought to clamp all kinds of restrictions on the larger units under the **MRTP Act**. While holding the check the unfair trade practices is logical all along, the diseconomies arising from low capacity had a striking effect on productivity, production and profit, as also on employment and technological innovation.

(ii) Liberal Investment Limits

The withdrawal of major restrictions under the MRTP Act and raising the investment limits for coming under the purview of the Act have in the meantime offered a relief to those units whose growth tendencies have been noticeable. Growth of size of business is a function of the size of investment called for by technology, size of the market, profit margins available and such other factors. The new industrial policy has promised to erase the distinction between the so-called public and private sector enterprises and the government has been hard pressed for funds so much so that shares of some of the profitable government enterprises are being disinvested. This being the scenario, the managerial challenges in the future regime are several.

(iii) Scenario Planning

First of all, management for the first time is being called upon to adopt scenario planning as a mechanism for withstanding the pressures that openness may create. Such scenarios may have to take into view, different possibilities and assumptions to make them realistic, so that management can choose from alternative scenarios the one best suited for future growth. Since there is no amber between the red and the green, a corporate enterprise either grows or decays, but cannot remain static. As such, the new conditions have made it as a duty that management indulged in futuristic exercises a lot more than before. Under insulated conditions, the options available for management were not many. Withdrawal of the permit and licence raj has opened up such possibilities.

(iv) Perceptual Mapping

Secondly, the new industrial policy has underlined the return of the free market forces so that the customer is treated as king and the survival or otherwise of a corporate enterprise would depend very much on how the customer views the performance of the enterprise. As such, it has become imperative for corporate management to adopt perceptual mapping as a major mechanism for knowing and judging how the present and future products and marketing efforts of the company are perceived by the society at large.

This is no small matter in so far as the perceptions of people change and there being keen competition in the market, the relative ranking of different products coming from different enterprises is a matter of fairly hard, sophisticated calculations. The customer being the king, his perceptions, howsoever inexact, would prevail. Product qualities, adequacy of supply timeliness of such supply, acceptable prices and such other related matters would, in fact, determine whether or not an enterprise can survive. Marketing orientation has remained an underdeveloped adjunct of Indian management; it has been so because the markets remained insulated from competition and demands outstripped supply many times more. Such hey-days of the last 40 years will be a matter of the past before long.

(v) Technological Upgradation

Thirdly technological upgradation will be a must in most lines of production and distribution. Technology may be divisible or indivisible. Where it is divisible, problems are not acute as the scale of operation can be modified with references to the technology available. However, where technology is indivisible, it will not only determine the capacity to be installed in a particular line of activity; even the inter-process balance, where processes are interlinked, will be difficult to attain. In such a case, utmost alacrity (eagerness) will be required to make the best out of a highly problematic situation that may be created by the fact that even when there is hundred per cent capacity utilization in the enterprise, individual processes may remain underutilized. Incidentally, enterprise capacity or overall capacity, is the capacity of the technically determined key process. In a chain-link system, it is the weakest link in the chain that actually determines the capacity of the chain. It will be a big problem for management to make both ends meet inasmuch as huge investments will be required to accommodate latest technology in a particular line, if market responses are not in consonance.

(vi) Market Creation

The growth in the size of the market is not just a matter to be watched from a distance. Enterprises will be required in future to create such markets and in this process, they will be called upon to undertake some social responsibilities, in the sense that the markets will not grow automatically; unless purchasing power is distributed in tune with the rate of progress sought to be achieved. Markets will shrink rather than expand in conditions where the consumers call the tune in terms of acceptability of products at offered prices. The problems are a lot more acute than anticipated; the choices that a consumer may be encouraged to exercise will relate to whether the products are of requisite quality, contain no hazards and are available at reasonable costs and prices. Agile manufacturing meets this challenge.

(viii) Innovation

Managers in an open economy will be expected to build up the scenario with reference not only to the handicaps, roadblocks or other constraints but also to the numerous opportunities that these difficulties may ultimately lead to; for, every problem in the managerial context is an opportunity, provided. Managers remain sensitive to the many avenues these problems open up.

Necessity being the mother of invention, history records the development of substitutes in conditions of acute scarcity of particular types of resources or where the costs of such resources spelt diseconomies of various sorts so that technological development such as Agile Manufacturing, Information Technology, Bioengineering, etc. could find a way by-passing the constraints posed by existing resources and uses.

(viii) Noise about Exit Policy

In the same context, it may also be mentioned that agro-industries will call for much larger attention than ever before. Even during the last few years, attention on agro-industries has grown on all sides. For example, sunflower oil, as a substitute for the traditional edible oil like coconut oil, groundnut oil and mustard oil, has received a boost in recent years. Innovation and creativity will hold the key to business success and in this process, unless the employment base is enlarged, even as a part of so-called 'private' enterprise responsibility, goods produced will not have many takers because of inadequate purchasing power. This is a clear prospect which can hardly be brushed aside in spite of all the noise created for 'exit policy'.

The 'thruster' organizations like agile manufacturers and enterprises expected to outnumber the 'sleepers' in future years will gain further in their efforts through increasing application of sophisticated management techniques and objective assessment of the business area for predictive and forecast exercise. The open market regime thus promises right prospects for the right group of managers and enterprises while the prospects may appear dim for those accustomed to resting on their oars.

12.1.12 Decision-Making in Agile Manufacturing

Making decisions is a fundamental process for any individual. In myriad circumstances he takes decisions regarding his basic needs (food, clothing, shelter, etc.), psychological needs (self-fulfilment) and social needs.

An organization in a sense is like a living organism and has to take many decisions to fulfil its organizational goals, namely, convert inputs into desirable outputs (products or services); to have a suitable structure (hierarchical or participative); to be able to communicate within; and to remain viable. In the entire process, decisions are taken at different levels to deal with men, material, machinery and money. Handling of information provides the basic ingredients in the decision-making process.

Decision-making process, whether undertaken by an individual or in an organization, can be clubbed under intelligence, design and choice activities, according to Prof. Herbert A. Simon. The first phase of the decision-making process can be called intelligence activity (borrowing the military meaning of intelligence) for searching the environment for information as well as conditions calling for decision. The second phase can be called design activity and would involve inventing, developing and analysing possible courses of action. The third phase can be called choice activity in order to select a particular course of action from those available.

As modern organizations grow more complex, information handling surprisingly goes up all the way. This is because the organizations have been transiting from pre-industrial to industrial civilization involving a changeover from agriculture to industry. Now they are well set to move from industrial to post-industrial civilization involving change from industry to services and leisure activities. There is a gradual sophistication of technology in these transitions and physical involvement with primitive technologies is becoming less important.

Organizations have therefore to deal with information in large volume. While men, material, money and machinery are still to be handled, they are increasingly coping with information-based decisions. No wonder the employees in modern organizations are called "knowledge workers".

Examples of knowledge workers can be legion. An employee of any banking organization is knowledge worker since he handles currency notes and banking instruments all of which are essentially pieces of paper containing certain information.

A doctor is a service organization is also a knowledge worker. He has to deal with around 7,000 names of diseases and about 70,000 medical terms in his day-to-day dealings and relate the patient information (symptoms) to the disease information (syndrome) that he knows, so that he can do the correct matching for diagnosis. Similarly, a therapy prescription is also a matching of information between the knowledge of disease and the knowledge of medicines.

A lawyer in a legal organization is another knowledge worker. For him the large body of constitutional provisions, statutes, laws and acts comprise the body of knowledge. This body has to be drawn upon to relate to a specific client, in addition to another large corpus of case law (precedents to be quoted).

Likewise, most workers in an industrial organization or a service organization also today deal with knowledge.

With the advent of industrial revolution, many decisions had to be taken regarding accounts keeping, product selling, order seeking, bill raising, salary decision, production watching and raw material procurement. As the revolution progressed, procedures evolved under which the above decision-making areas could be tackled in a routine manner. For instance, principles for double-entry bookkeeping; fixation of sales price and discount; maintenance of order books; procedure for invoicing; salary scales; standard output reports; and vendor fixation became available as regular procedures with pre-set logic; hence, the relegation of these routine decision-making areas with pre-set logic to the clerical levels.

A clerical system can be called a structured one, as all the rules and procedures are known in advance. This is today the decision implementation level. Complex non-routine and semi-routine decisions are taken elsewhere for smooth operation of the organizational activities. They can also be called the programmed decision system, as most of the routine procedures can be easily programmed before hand and automatically applied in practice.

Middle Management System

Proceeding from the same logic, the industrial revolution saw the emergence of managers who would be concerned with semi-routine decision-making. Going by our previous analogy, they would be deciding the cash-flow management; market research; order forecasts; credit management; personnel management; production planning and management (including finished goods and working progress inventory); and vendor analysis.

Interestingly the advent of computers and related technology is changing the practice and structure of middle management considerably. According to **Dr. Harold J. Leavitt and others**, the horizontal slice of the current organizations chart breaking into two, with the larger portion shrinking into a more highly programmed state and the smaller portion proliferating and rising to a level where more creative thinking is needed.

With the maturing of industrial revolution, this level has to take all the non-routine decisions, as outlined by **Dr. Leavitt.** Going by our charter of operations, the corresponding apex management decision-making would be in the areas of devising new financial instruments; industry profile and customer profile; planning of working capital and venture capital; personnel planning and skills forecasting; setting up new production lines and/or, deleting the existing ones, etc. Obviously, these decisions have far reaching implications and can be least structured.

Review of the above situation would make it obvious that any clerical system would need routine, quantitative, recurring, short-term, long-term data and internal data.

On the contrary, the managerial system (incorporating both middle and apex managements) would need non-routine, qualitative, non-recurring one time, long-term, global (incorporating many departments); and external data (with an impact from the outside economy).

As regards rules, the clerical system would involve structured rules (previously known and well-formulated) and certainty in logic (covering all possible alternatives). On the other hand, the managerial system would involve unstructured rules (not previously known and to be formulated on each occasion), uncertainty in logic (with increasing risk as alternatives are seldom known clearly or beforehand).

A clear goal of the managerial system is to expand the frontiers of decision-making both in numbers and complexity and push the routine decisions to the clerical levels for which automatic techniques are available. Another aim would be to improve the applications in more useful areas of management concern and gradually to high levels of managerial decision-making. In keeping with the above goals;

- The first task of a managerial system could be, reduction of uncertainty in decision-making (by using a non-routine information base). Often individual cases are put up for decision-making which do not have the backing of an objective basis in information. Especially in India uncertainty in logic makes for preponderant (greater) adhocism.
- The second task of a managerial system would be an intensive review and monitoring of the ongoing activities. While the clerical systems and the supervisory levels can carry on with set procedures and rules, the need for alert watch on the performance cannot be over-empahsized.
- The third task for a managerial system would be forward planning. The reuse of the past activities becomes of historical importance and such post-audit can be left to others. On the contrary, decision-making for the future plans is of paramount importance.
- The final task of a managerial system is control of operations. While detailed operations are of clerical and shop-floor concern, their overall control is a prime management process and has to be catered for in any managerial decision-making.

According to **Dr. D.K. Banwet**, size diversity, geographic dispersion, and complex production process, found in modern organizations demand sophisticated decision-making. The local and social environments are such that even the government record keeping and reporting requirements need the facilities of modern management techniques in decision-making. In particular, as industries become dependant on high technologies, they require greater precision in monitoring production processes and services. Finally, managers have quite often to perform multifunctional tasks and this involves an expanding role of management which was not required of their predecessors. All these make for an imperative need of managerial decision-making of the highest quality.

Benefits of managerial decision-making can be illustrated as:

- Improved allocation of resources under constrained conditions
- Rational utilization of human resources

- Increased efficiency of output in terms of quality, reliability and productivity
- High quality service to customers
- Timely achievement of targets and schedules in terms of overall profit and reduction in time and cost.

Managers have always had a system to provide themselves the information for their decision-making. Some of the oldest known examples of written language are on clay tables that preserved the transactions of the ancient Mediterranean merchants. When these records are deciphered by archaeologists, their similarity to modern management information is peculiar; there are the equivalents of inventory records, accounting statements, receivables and even what appears to be production records for piece-rate workers. Perhaps modern managers have to discover these information details to help in their decision-making.

12.1.13 'Workforce' Participation in Agile Manufacturing

The workers' participation in management is a concept much talked about but little understood and implemented. The frequencies and the manners of its uses not only by all other shades and hues of people have resulted in the loss of its precise meaning. The vagueness that has gathered round this concept has created confusion in comprehension of its concise connotation (meaning). In order, therefore, that ideas do not get tied up in knots, it would be fair to clarify the concept in the way it is grasped. In the ultimate analysis, the workers' participation in management, implies, taking part by the workers in the decision-making apparatus of the management. Management can be considered as an integrated system of decision making with the avowed (declared) aim of realizing the set goals. This decision may pertain to the areas of organizing, planning, coordinating, controlling, directing, motivating and innovating which in a nutshell constitute the corpus (written down procedures) of management process.

Workers' participation in management can thus be conceived as an active, willing and enlightened involvement of the workers in the decision-making machinery of the management with a view to achieving effectively the organizational goals. Devoid of this meaning, the concept hardly seems to hold any significance or relevance for the organization which is basically a corporate enterprise or institution.

This decision-making process is a highly complicated and sophisticated process, involving varied skills, articulation (clear communication), education, training and enlightenment on the part of the decision-makers and as such seems to be almost out of reach of the sizeable number of Indian workers drawn largely from the agrarian background, deficient in the required education and skill. This, however, should not lead us to a hasty conclusion that the stage is not at all ripe for Indian workers to participate in the intricate decision-making process of the management. Nevertheless, it certainly throws a mild hint that we require to tread slowly and gradually in this direction, lest the result will not only be frustrating but even leading to utter failure. The experiment, if not launched with due deliberation and diligence, is likely to mitigate the realization of the dream of democratization of Indian industries. What is needed, therefore, is to start implementation of this concept at the lower echelon of management in the first phase and thus to raise it gradually to the higher decision-making levels, reaching ultimately the Board of Directors level and the corporate level.

Workers' participation in management should normally pass through the different stage of its growth starting from sharing of information with the workers, consultation with some important matters, collaboration with them in certain selected areas and thereafter, their participation in the decision-making machinery of the management. This gradual growth will generate a climate of confidence in workers as well as lead in an atmosphere of participative management in the Indian industries.

Workers' participation in management is an extension of democratic principles in the industrial endeavour. The concept has grown with the growth of democratic consciousness in our country, though it has gained currency recently owing to its inclusion in the 20-point programme where it has been accepted as a part of our national economic programme. It has since then ceased to be mere craze or fashion and assumed a tone of seriousness, inviting all our efforts and endeavours to give it a real shape, to translate this principle into actual practice, to make it a way of life and to cultivate this culture in the industrial management.

We must take a vow that once our leaders, makers of the nation's destiny, have lighted this lamp, we must keep its amber burning aloft and would not shut our eyes with any sign of defeatism and escapism, with any fear of sadness and disspiritedness.

Vivekananda has very rightly said that "one ounce of practice is worth hundred tonnes of mere talk". We have really talked a lot about this participation but have actually practiced it little. The time is now not only to give it a lip service but to make it a way of life by cultivating a cooperative culture and fostering a spirit of partnership 'in the management of business'.

Three Trends

The growth of this concept can be conveniently conceived if you take a look at the three trends that have emerged since the dawn of industrial revolution regarding the relationship between the management and workers.

- The first concept seems to veer round to master-servant relationship especially during the proprietary management, treating workers merely as domestic servants, hiring and firing them at sweet will, whims and fancies. It was largely believed that workers either require a kick behind or a carrot in front to make them work.
- With the growth of social awakening and the rise of professional management with the passage of time, this concept of master and servant relationship gave way to employer-employee relationship wherein the relationship was regulated and governed more by contractual obligations, being either legalistic or quasi-legalistic in nature rather than being merely unpredictable and whimsical. The gap between the workers and the management still remained intact, the warmth of relationship was still a far cry; the dichotomy (division/contrast) was still too deep. The real relationship which is the foundation and fountain of the partnership spirit was yet to be established. The relationship remained more or less mechanical and dissipative rather than cohesive, integrative and organizational. This state of affairs compelled for a better and more cohesive and cordial relationship so that the objectives of the corporate enterprise where every work has an element of interdependence can be achieved efficiently and effectively. This orchestration in relationship between the

management and workers has to be set up in order to avoid any discord and disharmony resulting in the loss of productivity as well as to ensure a sound organizational health and enhance the economic competence of the enterprise thereby contributing to the growth of national economy for the amelioration (betterment) of the lot of teeming millions, for raising their standards of living.

- This thinking has given impetus to the emergence of the relationship of partnership between the management and workers where both the parties are considered to be the co-partners in the industrial games. This concept is to be fostered not as an article of fashion but as an article of faith—a faith emanating from the depth of the heart that the democratic way of life, the participative style in interpersonal dynamics, is the best and the most effective way for the economic salvation of the nation. Thus, the upsurge of the democratic ideas and ideals, coupled with the problem-solving approach, beliefs and ideology cherished by the nation, has provided a fillip to the implementation of this concept.

The question arises as to why we talk of this concept so shallowly. The answer lies in the urge to participate in the management process which is linked with the inner psychological springs motivating human beings to give off their best to the professions they are in, as they derive the status and dignity from such involvement. The participation satisfies the ego needs of the workers, elevates their status, instil in them the organizations and makes them committed to the organizational goals. It fosters the feeling of oneness with the management which results in the acceleration of their efficiency and effectiveness at work. The decision in which the worker participates, the affairs in which he is consulted, the area in which his collaboration is called for become his own and once he owns them, he spares no pains to implement them as effectively as he can. If this rapport is set up, the task, however, arduous, becomes easy to accomplish.

Status

From the inception of the industrial revolution, the thinkers have been seized with the idea of enhancing workers' status. Some have even gone to the extent of advocating equalization of power between the management and workers, which seems to be an ill-conceived notion, neither practicable nor feasible. Power is derived from various sources, both from within and without, and our control being very feeble on these sources, the talk of equalization of power in the creation of the providence who has not made even two leaves exactly equal is a wild-goose chase, a mere piece of imagination. Nevertheless, there is ample need for sharing power and bridging the gap between haves and haves-not. Thinkers like August Comte and Robert Owen have advocated the participation for the sake of distributive social justice. Karl Marx and Engels favour the complete control of the national economy by the working class. Scholars like Cole and Weble have however, felt that only participation of workers in management process would suffice for enhancing the status of the workers and increasing their commitment to the organizational goals.

Inspired by these ideas, different nations of the world have translated this concept into practice keeping their national cultures, ethos, mores (customs), ideals, beliefs and ideologies in view. The nature and the extent of the involvement of the workers in management process, therefore, differ widely from country to country.

The co-management of **Yugoslavia**, co-determination of **Germany**, grass-root level participation of **Scandinavian** countries, the proletarian dictatorship of **Russia** and the joint consultation of the **United Kingdom** are some of the examples of the workers' participation in management in their characteristic, inimitable styles. The American workers are generally not bothered about the participation in management as they hold the view that the managing of business is the business of management. The affluent working class of American industries derive status and dignity and share the gains of the industry by resolving the conflict through collective bargaining machinery. The strong, stable and representative unions in the permissive society of America speak with management on terms of equality and strength, though this process is basically a give and take process and not participative in the real sense.

Since the culture, beliefs and ideology of the nation determine the nature and extent of participation, style and strategy of participation, we must bear in mind that whereas we may draw inspiration from the other countries of the globe in giving shape to the participation in Indian management, any blind imitation in this regard is not only costly but also dangerous. Our system of participation must be modelled on our own national beliefs and ideologies. Our system and style must be congruent or relevant to our national ethos and must be suited to our soil.

Inner Leadership

In the context of workers' participation in management, only work-based and worker-oriented people should be the leaders and not the politicians or external leaders, (who are for self interested in gaining leadership) though they profess to be the saviours of working class. The leadership must evolve from within the working class keeping the interest of the work as supreme in their mind and not imposed from without. Hypocritic and masked leadership are not the need of the hour but a work-based internal, considerate and enlightened leadership. However, as the conditions stand today, our industries are highly deficient in work-based and work-oriented leadership and there is a serious need to bridge this gap if this concept is to be given a fair deal. When externals held the reign of leadership, it may be called as intruders' participation in management or the external participation in the internal affairs of the management which is not a symptom of strength but of weakness of the organization. Only with the internal, work-based leadership, a culture conducive to participation can be created. This experiment, of course, calls for change of organizational structure and strategies. The gap between the management and the workers will have to be narrowed down. The hard and fast line of distinction between the two classes has to be eroded. The rigidity of the existing structure has to be replaced by an adaptable and flexible structure which may pave the way for participative culture. A stable, workable structure has to be devised and designed for the furtherance of this cult.

The participative culture demands a sacrifice on prerogative centred approach of management for its substitution by the problem-solving approach through participation. This shift entails sacrifice of a good deal of power and prowess hitherto enjoyed exclusively by the management elite and also tremendous preparation on the part of the working class to enable them to participate the decisions on management issues.

Participative culture is basically a democratic culture as against controlling and dictatorship cult. **E. M. Forster** in his "Two cheers for democracy" has clearly stated "so two cheers for democracy

—one because it admits variety and two because it permits criticism". In this statement lies the typical example of democratic culture which has to be imbibed in order to foster a participative spirit in managing the business. Admission of variety implies acceptance of the individual differences, appreciating the individual views and permission of criticism entails appreciation and tolerance of the dissent. In the practice of participation, these two democratic virtues have to be cultivated, cared for and commended. Without adopting such psychological spirit of mind, no participative machinery will yield desired results.

There are two main suspicions raised against the success of the participative culture in our country :

(i) One is that either workers or managers or both are not ready for it.
(ii) The other is that the atmosphere of the country is not ripe for implementation of this concept in industrial management.

Conditions

There is no denying the fact that giving certain conditions conducive to implementation of this concept, its effectiveness and success are assured. These conditions mainly are the following:

(a) Workers must be satisfied with the level of wages and other fringe benefits;
(b) Trade unions must be woven out of the warp and woof of internal working force with work-based bias. There must be a strong, stable and representative union owing the allegiance of a sizable number of workforce. They must be prepared to subordinate to a great extent their political ideologies to industrial considerations;
(c) Management must be willing to shake off their paternalistic and autocratic attitude for a democratic outlook, tone and temper;
(d) There must be a considerable period of industrial peace in the enterprise and the country at large; and
(e) The enterprise must have a reasonably long period of economic stability, and so on and so forth.

Though the presence of these factors in any enterprise is bound to have its unpleasant impact on effective implementation of this concept, nevertheless, we should not stretch it too far that unless these factors are fully present, the concept of participation should not be given a trial. Such excuse seems to be an offshoot of a lukewarm and complacent attitude, which has to be scrupulously avoided. Democracy is a self-educating process, and once this democratic style, this participative culture is cultivated even by way of rituals, in course of time, it is likely to be engraved in the organizational ethos. Though suitable climate or atmosphere generally generate the effective structure and style of the organization but sometimes the structure and style also induce the conducive climate and adaptable atmosphere. As such, the deficiency in the climate of the organization should not very much deter us from launching the participative schemes.

All the objections or suspicions that are received from the management side are essentially based on the feeling that the workers' participation in management will take away some of their prerogatives which they have monopolized and hugged as precious treasures, so far. This initial hesitancy, however, will be overcome by moulding the attitude in terms of problem-solving approach rather than by the concentration on personal powers and prowess. No doubt, the concept is revolutionary, demanding a sacrifice of the managerial prerogatives.

The unions, on the other hand, feel that participation will reduce the powers of the unions and also that workers by joining in the management process will compromise their freedom to criticize and press legitimate demands. There is no doubt that by joining in the management process the workers bind themselves to abide by the decisions in which they have themselves participated but this is one of the basic purposes for which participation is planned. These suspicions basically are raised by the external leaders who are generally not work-based and worker-oriented. If the leadership is appropriate and enlightened and the commitment to the organizational goals is ensured, these suspicious do not stand.

Last but not the least, success of the workers' participation in management should not be measured in terms of the number of decisions mutually agreed upon and the absence of disagreements but by the extent of feeling generated in the workers that they are also the part and parcel of the dynamic process of management and not merely the cogs in the machines. The main purpose is to spread a real feeling of partnership among all those who contribute their mite for industry and the national economy at large.

In the end, it may be pointed out that if the vibrancy and viability in national economy has to be attained, the industrialization has to prosper in India, for which the management and the workers must work hand in gloves with each other and this can be possible only if all-out efforts are made to promote the participative culture in industries.

12.1.14 Empowering for Agile Management

The general concern in corporate circles is how to prepare organizations and people for the global market economy. How to train people to respond quickly to the fast changing environment is the question on everyone's lips. These aspects have been often discussed at the annual convention of **ISTD (Indian Society for Training and Development).**

While training can help in imparting knowledge and skills, what really enables one to perform is the initiative that the person himself takes. A new concept now gaining ground in this context is 'Empowerment'. What this actually means is that a person's willingness to act depends upon whether he feels empowered to do so. The source of this empowering may be from within or from his superiors.

Quite often when people complain about not having enough powers to act, the question arises as to what holds them back when powers are there but action is missing! Obviously, it is not only the actual powers but the 'felt' powers that make people act. How to generate this feeling is the focus of the concept of empowerment. This concept is vital today as quick action is the key of success, when one is dealing on the global level. People cannot wait for decisions the same way they used to while dealing only with domestic market, when time was of little consequence. This had habituated them to wait long time for decisions to emerge from the top management only!

The empowered manager shifts the power for employees to personally take responsibility for the success of the business, if it is to survive and prosper. Middle managers especially are the ones who feel helpless today to influence the events. They feel completely hemmed in by the trade union power on the one side and the top management's perceived authority on the other. Yet it is they, by virtue of the operational skills they possess, who can really transform the organizations.

How to realize this energy of middle management? This is the crux of the problem. The dilemma of the middle managers is that if they choose to act fast, will they be supported or let down? There are two aspects to this:

(i) Part of this answer has to come from oneself by being courageous and avoiding the path of being courageous and avoiding the path of being merely cautious and all the time looking up for support.
(ii) The other aspect is the facilitating organizational climate, to foster the ability to take calculated risks to achieve growth, not getting into a rut and remaining stagnant.

Risk of New Initiatives

Traditionally, since dependence is very much fostered, exercising autonomy even if one has to consider the choices that shape one's own work environment. There is always an element of risk involved in not being a conformist and choosing new paths. But one has to also recognize that some support has to be mobilized to steer new ideas through and get them accepted by the powers that be.

Choosing an entrepreneurial path as against a bureaucratic one is not always easy. The fear of failure is a powerful deterrent as the common feeling is that mistakes are punished more severely than achievements are rewarded. On the other hand, the bureaucratic belief in policies and procedures becomes a source of security.

In this context, what is the way out? An autonomous or entrepreneurial mindset means treating the business as if it is one's own and taking the necessary actions in time. This is a better alternative than to wait for instructions from above and depend upon the bureaucratic channels all the time.

In empowering oneself to act, what is important are the following four steps.

(i) Creating a vision of greatness for the unit.
(ii) Identifying and managing allies, adversaries and fence-sitters.
(iii) Understanding and letting go of one's own wish for dependence.
(iv) Discovering the courage required to live out the vision.

Peter Block suggests that one should begin with one's own customers and that too with the internal ones. He suggests the following questions:

- If our internal customers were the only customers, how would we treat them?
- How can we use their frustration and disappointment with us as a learning experience to teach us how to improve our way of doing business?
- How do we deliver bad news to our users? Unfulfilled promises—instead of blaming it on other, take responsibility for our mistakes?
- How to handle situations in which our customers let us down? Part of our function is to help customers be more effective in their own way of operating, showing them ways to work better with us.

Critical Skills

It is not only important to have a vision; it is equally important to build support for it. In this respect, the critical skills called for are negotiating an agreement and negotiating trust. The first step in developing a practical strategy is to identify who among the customers, bosses, subordinates and bystanders, one needs for the success of one's ventures. The approach is to work through three steps:

(i) Exchange vision, purpose or goals.
(ii) Affirm or negotiate agreement.
(iii) Affirm or negotiate trust.

Figure 12.1 illustrates the possible combinations.

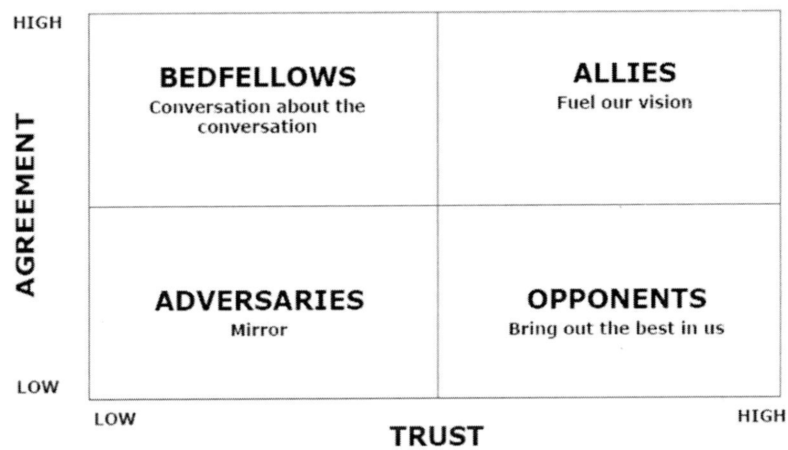

Figure 12.1: Matrix showing allies (who fuel our vision) and opponents (with whom limited agreement is possible).

The above matrix tells us that we need a healthy mix of:

- **Allies** who fuel our vision, and
- **Opponents** who bring out the best in us.

In respect of allies, it is always advisable to ask for their support. In the case of opponents, only a limited agreement may be possible. The most difficult problem is with adversaries where containment strategies will have to be worked out. In the case of bed-fellows, "convenience" arrangements will have to be worked out. These are termed "positive political skills" by **Peter Block**.

Ultimate objective of empowerment of oneself is balancing autonomy and dependence so that one develops practical strategies to live with realities and one has peace in the environment that one works in.

12.1.15 Teamwork in Agile Management

The quest for a new work culture is a phenomenon, felt not only in India but the world over. To create such a culture in organizations, it is essential to formulate desired success experiences among employees, which would help in developing the necessary attitudes and value systems. This new culture can be brought about by a change in the management style. 'Team spirit' is essential to achieve new heights in workers' performance. There is not only a need to develop achieving teams in the framework of formal organizations, but also to nurture voluntary teams for many innovative tasks.

The call of the hour in India is to develop a new work culture and ethos. In the United States, the business community is seriously attempting to adopt a style of management quite different from the historically evolved one. The focus is on **Theory 'Z'**, otherwise known as **'The art of Japanese management'**. It is based on an organizational culture of loving care to all employees, guaranteed lifetime employment and management, a team approach in problem solving and meeting challenges, an opportunity for all to use talent and ability, an atmosphere of growth and stability, etc., ought to be the guiding principles of this new culture.

New Styles

The primary task is exposure of management to the new styles. Managers are usually accustomed to conventional styles where heroic models of leadership are prevalent. Exposure by theory is of no avail. Retraining managers, especially the middle level and front line, to adapt their leadership behaviour in order to develop the perceived organization culture and work ethos will help to accelerate the process of new culture development.

Attitudes, values and beliefs are developed by people through the needs and events experienced by them. When a person feels a need or faces some event it generates emotions and feelings which propel action(s)/challenging goal will have to be set in tune with the organization goal. The team must be capable of handling problems and taking decisions such that each member shares the responsibility willingly. The style of management leadership, means achieving team developer leadership, relying on developing individuals in the team through mutual support and creative participation.

Achieving team developers have imbibed their philosophy from great leaders who believed in the abilities of people. Every one strives for a sense of achievement. People with diverse backgrounds and interests can and want to work together and contribute to achieving a common goal. Great leaders not only bring about group involvement in various programmes and movements but also make the members capable of self management.

Formal vs. Informal Organizations

Management pundits generally focus on two types of organizations – formal and informal. In a formal organization, the authority flows from the top while in an informal set up, it is bestowed on leaders by their followers.

- In a formal organization, every employee has a specific role to play. He or she is burdened with routine duties and assigned responsibilities. There is not much scope for creativity, and inevitably, boredom and monotony set in. The formal work teams confine their interpersonal relationship to the members of their section or department.

- The informal organization is considered a useful instrument. Today, with the popularity of quality circles, problem-solving teams, project teams, etc., organizations are slowly realizing the importance of another type of organization – voluntary teams. Quality circles, task groups, value engineering groups, sports clubs, etc., are formed mostly on the basis of voluntary membership. These groups bring out the creative abilities of employees and given them the opportunity to use their talents and capabilities, which otherwise would not be put to use in their regular jobs.

Voluntary Efforts Effective

Japanese management encourages informal relationship between supervisors and subordinates making them feel at home. Quality circles is one of the numerous voluntary teams cultivated in the Japanese industry. American industries are also taking pains to cultivate voluntary groups like quality circles, project teams, think tanks, mega teams, etc. Whenever a new method or technique was evolved it was their practice to confine it to a designated department as a formalized function. When value engineering came up and **Statistical Quality Control (SQC)** got under way, these became departmental functions. But in Japan, many functions which have scope for universal application get into the shape of movements in organizations and they are spearheaded by voluntary teams formed with the consent and interest of the members of the team. These teams cut across the normal workplaces and departments and their functioning even stretches outside the normal working hours.

Creative voluntary teams can be cultivated by organizations as is done for Quality Circles. However, they involve a strong faith of the top management that every employee has creative talent and if the opportunities and environment are provided, he would use it for the benefit of the organization. The top management should support and encourage the employees to form and manage their own teams voluntarily.

Case Study

A certain multinational corporation with its factory at Mumbai was facing several problems – industrial relations as well as financial. It was not able to carry out any changes in the production system and was running under a loss for two consecutive years. At that time the management felt the need to train all their managers, from supervisors to senior management levels in behavioural science. They called the **National Productivity Council (NPC)** to design and help in conducting their **HRD** activities. The training programme concentrated on problem solving and decision-making.

After a series of these programmes, the training department with the assistance of NPC made it an opportunity for OD. The mechanism used for OD was forming voluntary teams. These were called ‘mega teams’ and comprised members from various levels of management and departments, including some leading workers from the shop floor. Each team took up an important problem according to its own interest. After initial teething troubles, these mega teams became well-knit and within a year brought about several changes – organizational and technical. The organization saved more than Rs. 90 lakhs and helped the unit recover from its sickness.

12.1.16 Leadership in Management

Leadership is one of the most talked about subject in industry and defence services. Academic and behavioural scientists have been working constantly to upgrade the approach to leadership and help organizations to make the best use of their leaders. The basic definition of leadership "the ability to influence a group of persons in achieving an organizational goal effectively with maximum results" remains valid. Ultimately the handling of human resources is the key to success for managers besides massive material resources.

Management is the art of planning, organizing, directing and controlling the material resources. Leadership training can be imparted to improve the personal skills and ultimately it depends to a great extent on attitude, behaviour, perceptions and most important concern for the men working.

Winston Churchill, Rommel, and Mahatma Gandhi provide a few examples of those who brought glory in spite of poor resources and adverse situations. The stakes are too high to ignore the art of leadership in the present environment.

In defence services, few decades ago, much emphasis was laid on traits such as courage, tenacity, endurance, honesty and foresight gained much importance due to the service environment, its selection process and the mythology for the training of leaders. Popular notions of public schools and success stories of personalities such as **Nelson and Wellington** lent much aura to the personal traits of leaders. However, subsequent studies have proved that there is no overall correlation between the traits and the performance in the field. As missions became more complex, attempts were made by social scientists and researchers to seek new answers to the age old question. How effective is leadership ensured?

The **Ohio University** studies based on attitudes and behaviour found that the weightage given to task accomplishment and concern for men varied with leader. They were plotted to form the two dimension models. It was postulated that a combination of high task accomplishment with equal concern for men could form the pinnacle of a good leader.

Leadership Styles

The managerial grid concept made popular by **Robert Blake and Mouton** further improved the Ohio model by depicting various leadership styles that have become the subject of many studies. Emphasis in this model shifted to the leadership styles which can bring out the best in a leader. Leadership style can range from the highly autocratic to country club which uses the minimum authority.

Refinements to this study were attempted by **Paul Hersey and Kenneth Blanchard** who added the third dimension, that is the environment. This study emphasizes the well known observation that the leadership styles cannot be uniform for all situations. A crisis situation such as floods, accidents or gas tragedy will demand from the leader, an authoritarian attitude to propel a group into action. Normal problems in industry must be given due deliberation and consultation before a decision is taken.

With an understanding of the above arguments, it is now easy for a manager to understand and implement some of the guidelines laid down below. Needless to say that the leaders must possess the common traits such as knowledge, integrity, intellectual honesty, confidence and tenacity in order to lead his men.

(a) **Verbal communication:** Once in a week or fortnight, talk to all the staff informally and keep them informed of your plan of action. At the same time, elicit from them their individual problems and fears. Such enlarged open house sessions in an atmosphere of informality could pay handsome dividends. **Field Marshal Montgomery** used to go round and speak from a jeep during his command of the **Eight army in the African deserts**.

(b) **Visits:** Too many managers get tied down to a cabin immersed in their paper work. An occasional visit to the shop floor or service department would enable the leader to listen and find out the personal problems of those who are too shy to confine. If this could be followed by suitable action to help the staff, the net pay off will be tremendous. In a crisis situation, a leader must not hesitate to shed his coat and get down to help with his bare hands.

(c) **Canteens and recreation:** In spite of a busy schedule, the manager must find some time to visit the corporate canteen, sports club or recreation room to see for himself the items needing rectification or improvement. Ideally, these should be surprise visits to obtain first hand information on important aspects of individual relations and welfare.

(d) **Sickness and death:** A good manager must be able to know immediately the serious cases of illness or disablement. An impromptu visit to the hospital or even to the residence of the ailing staff will bring a tremendous sense of confidence in the lower ranks. It will reassure the juniors that the leader is a man who cares for human beings. **Mother Teresa** is a fine example of a social leader who displays this attitude and behaviour.

(e) **Social interaction:** To foster human relationships, the manager must sponsor events such as sports gathering, annual day and children's functions in which he and his family can have access to the families of his subordinates. Such occasions could be used to instil in the dependents a sense of belonging.

(f) **Appreciation:** A work of appreciation at the appropriate time for a job well done can create a lasting impression on the subordinates who have the need for recognition and affiliation. In addition to the material needs, an individual who has reached the middle level of his career, craves for attention and compliments. During World War II, **Winston Churchill** found time in his busy schedule to write a line of praise to the subordinate commanders who were located in various parts of Africa and Asia.

(g) **Credibility:** A leader must prove his credibility by total implementation of his promises and assurances. No amount of talking can restore a leader's stature if he fails to follow up his words. Montgomery told his soldiers that throw out Germans cannot commence unless the British army was provided with all the material recourses. He stood by his words and consequently he reaped the benefits of a magnificent victory.

In today's environment when history is beset with problems of productivity and financial stringency, the human resource constitutes the most important factor. Even industrial giants have realized that management of material cannot be isolated from the fine art of leadership.

12.1.17 TQM in Agile Management

The term which is now on everyone's lips is exports. The question uppermost in the minds of the captains of industry is how soon and how rapidly they can increase exports.

Quality is the key to this whole effort and the need of the hour is TQM. It is therefore appropriate that this aspect has been discussed many times.

Since production of an item in an organization involving a series of processes is like a relay race, the performance in the last lap alone cannot win the race. Every process from the beginning to the end has to be correct to produce a quality product. Therefore, the people involved in each process have to treat those in the next process as their customers. Thus, in addition to the external customer, every activity in the factory has an internal customer. The service to the internal customer is the one which will help gear up the organization to deliver finally the required service to the customer. Thus, for example, if the marketing department of an organization deals with the external customer, its commitments are to be backed up by other departments for whom marketing itself will become the customer.

Internal Customers

A lot of lip service has been given to this concept, but in reality the internal customers are not taken seriously. Instead, if a defect is pointed out there is mutual acquisition and mud-slinging at each other. What is needed is not only expression of vague intensions, but clear cut spelling out of what exactly are these internal service parameters. It has been the author's experience that such an exercise conducted seriously requires all the managers to sit together and spend a couple of days in spelling them out. For example, the design department will have to spell out the services required by the manufacturing division from design and how design will actually fulfil it.

Similarly, the marketing division will have to provide field trial data to the design department to let them know how the products developed by them actually function in the field. Thus, a full-fledged exercise of all the departments can help crystallize clearly the services required from each and the quality indicators that will throw light on whether these services rendered will help meet the ultimate quality specifications of the customer.

Once these quality parameters are spelt out, an attempt can be made to evaluate where the gaps are and what steps are needed to bridge these gaps. Once the gaps are identified, then the right tools must be selected for addressing the problems. There are seven quality tools which have been popular.

(i) Pareto Chart. (Fig. 12.2 a & b)
(ii) Check sheets (Fig. 12.2 c)
(iii) Cause and Effect Diagram (Fig. 12.2 d).

(iv) Scatter Diagram (Fig. 12.2e).

(v) Control Charts and
(vi) Histograms (Fig. 12.2 f)

(viii) Graphs (Fig. 12.2 g)

(i) Pareto Chart Examples: Figure 12.2a (Example 1) shows how many customer complaints were received in each of five categories.

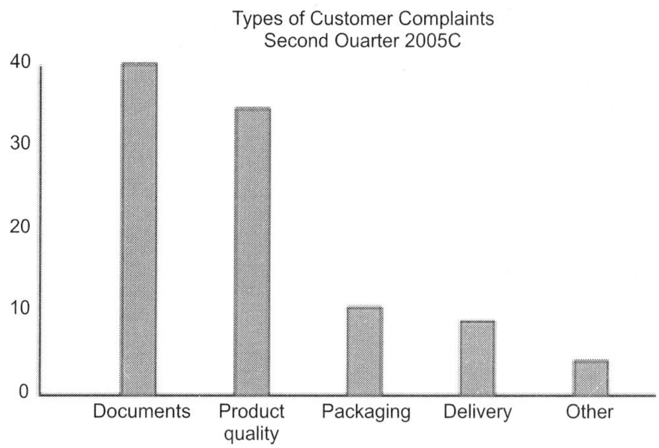

FIGURE 12.2a

FIGURE 12.2b

Figure 12.2b (Example 2) takes the largest category, "documents," from Example 1, breaks it down into six categories of document-related complaints, and shows cumulative values.

If all complaints cause equal distress to the customer, working on eliminating document-related complaints would have the most impact, and of those, working on quality certificates should be most fruitful.

(ii) Check Sheet Example: The figure below shows a check sheet used to collect data on telephone interruptions. The tick marks were added as data was collected over several weeks.

Telephone Interruptions

Reason	Day					
	Mon	Tues	Wed	Thurs	Fri	Total
Wrong number	卌	II	I	卌	卌 II	20
Info request	II	II	II	II	II	10
Boss	卌	II	卌 II	I	IIII	19
Total	12	6	10	8	13	49

FIGURE 12.2c: Check sheet sample.

(iii) Cause and Effect (Fish bone) diagram.

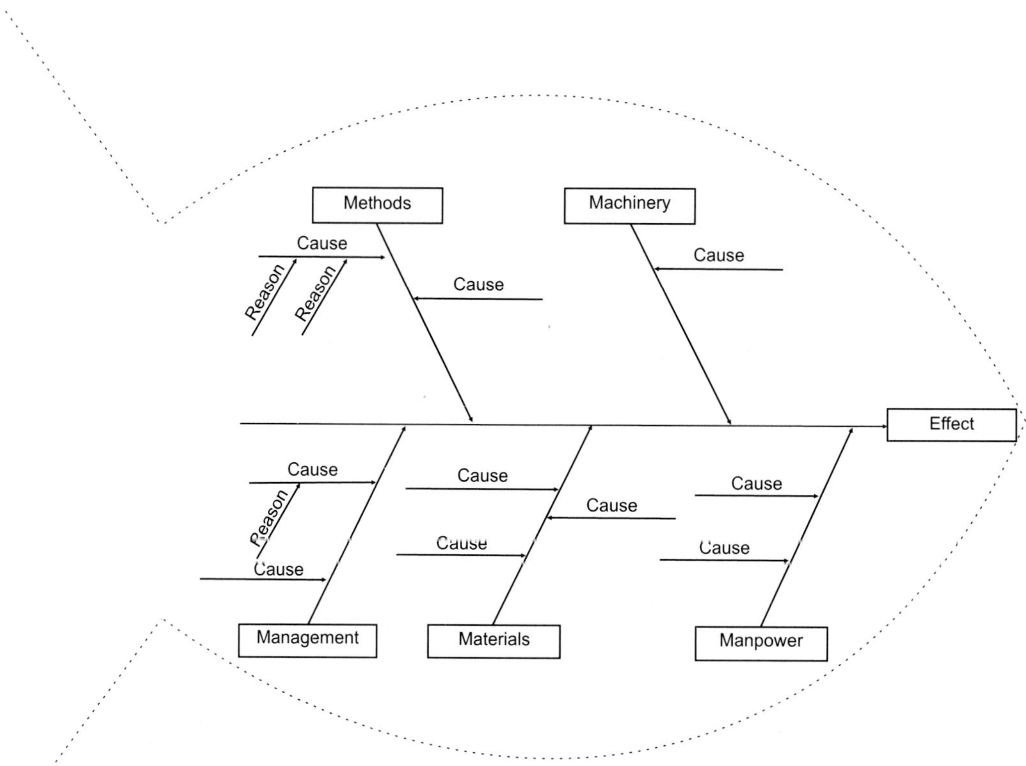

FIGURE 12.2d: Cause and effect (fish bone) diagram.

(iv) Scatter Diagram

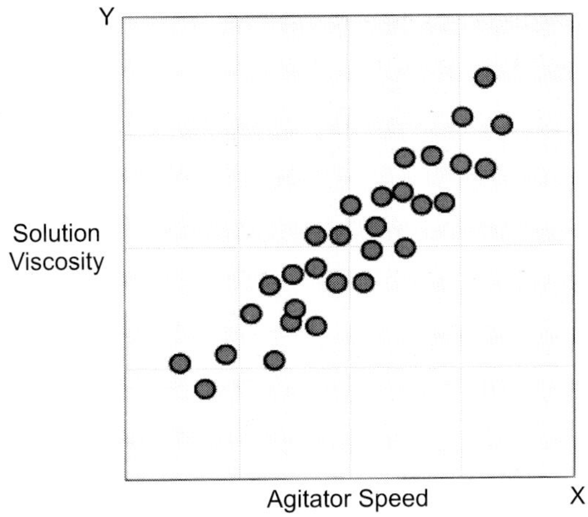

FIGURE 12.2e: Variation in solution viscosity for different agitator speed at different intervals of time (scatter diagram).

(v) Control Charts & Histograms: The general approach to on-line quality control is straightforward. We simply extract samples of a certain size from the ongoing production process. We then produce line charts of the variability in those samples, and consider their closeness to target specifications. If a trend emerges in those lines, or if samples fall outside pre-specified limits, then we declare the process to be out of control and take action to find the cause of the problem. These types of charts are sometimes also referred to as Shewhart control charts (named after W. A. Shewhart who is generally credited as being the first to introduce these methods; see Shewhart, 1931).

Interpreting the chart. The most standard display actually contains two charts (and two histograms); one is called an X-bar chart, the other is called an R chart.

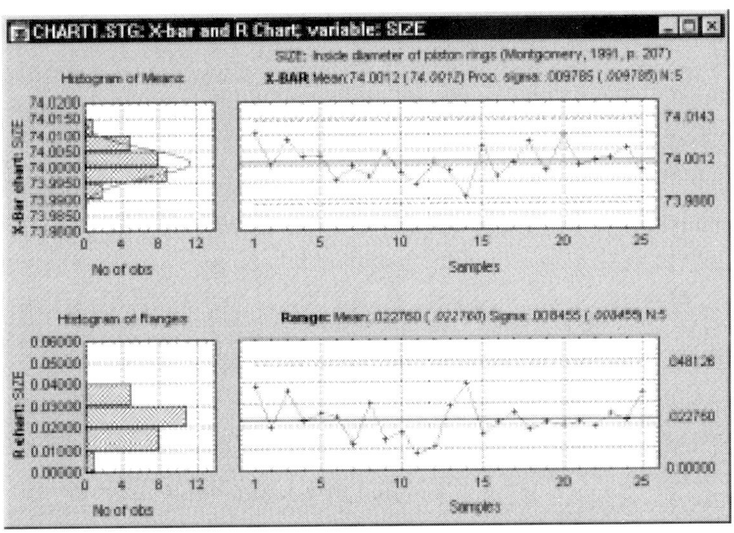

FIGURE 12.2f: Control charts and histograms.

These are all statistical in nature and require data for analysis. The application of these tools has not been that widespread in India industry as, for example, in Japan, it is this widespread use by all levels from top to bottom in industries that has given them the quality edge.

(vi) Bar Graphs: Perhaps the most common type of graph is called a bar graph. Bar graphs come in handy when your series is, say, 10 data points or less, or when you want to emphasize the magnitude of your data (as expressed by the height of the bars).

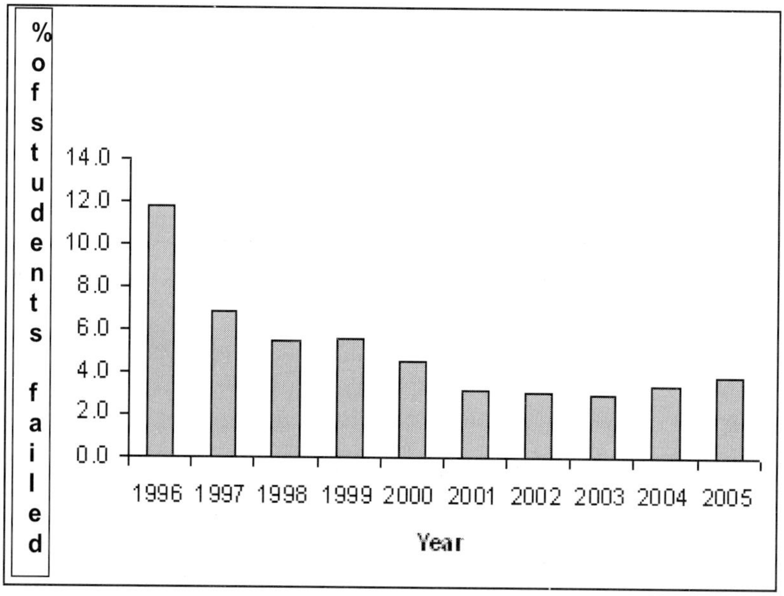

FIGURE 12.2g: % of students failed during years.

The New Seven Quality Tools (Fig 12.3 a to 12.3 g)

In addition, there are seven new tools which are "language driven", for use by managers, to define and diagnose problems that are not easily quantifiable. These tools also help translate the customer requirements into actual product requirements. These new tools are:

(i) Relations Diagram – This clarifies the interrelations in a complex situation involving many related factors and serves to clarify the cause-and-effect relationships among these factors. For example, speedy response to the customer can be analyzed through this tool.

In many problem situations, there are multiple complex relationships between the different elements of the problem, which cannot be organized into familiar structures such as hierarchies or matrices. The Relations Diagram addresses these situations by showing relationships between items with a network of boxes and arrows. The most common use of the Relations Diagram is to show the relationship between one or more problems and their causes, although it can also be used to show any complex relationship between problem elements, such as information flow within a process.

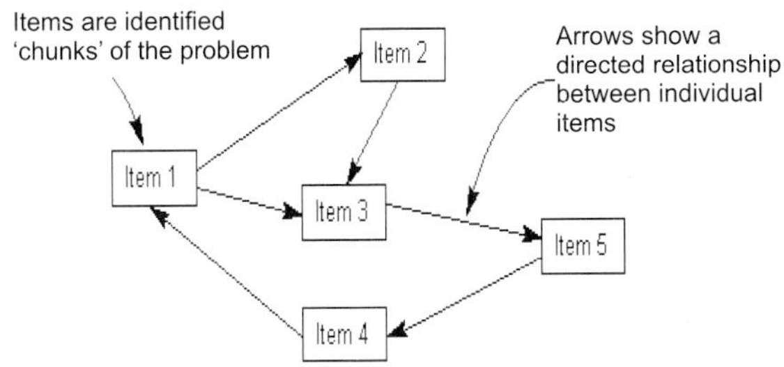

FIGURE 12.3a: Relations diagram.

(ii) Affinity Diagram – This is for grouping the factors which are of an allied nature. This is essentially a brainstorming method.

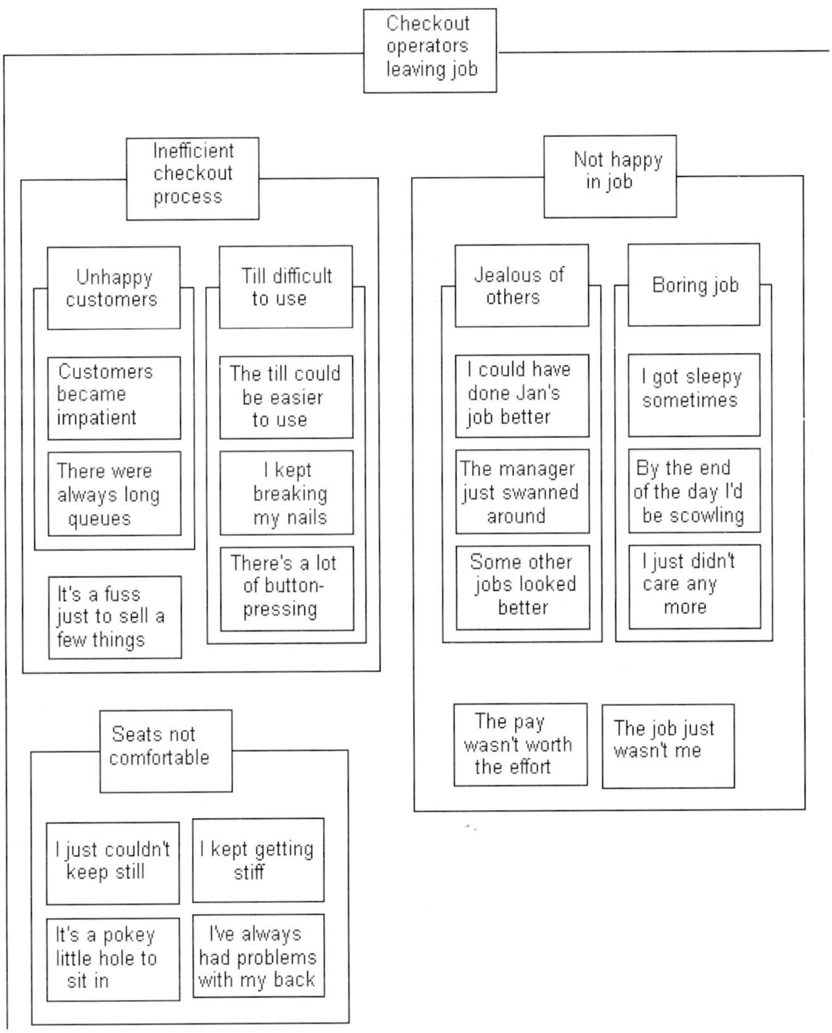

FIGURE 12.3b: Affinity diagram.

Example: The personnel department of a food supermarket chain store identified a high resignation rate of good checkout staff. There was information available from exit interviews about their reasons for leaving, but this was disorganized and there was no clear area that they felt they could address. They decided to use the KJ method and an affinity diagram to try to better understand why these people were leaving (see as illustrated). As a result, the checkout process was investigated further and eventually completely redesigned. This included a redesigned booth and hourly breaks for operators. Consequently, there were significantly fewer leavers (and as a bonus, customer satisfaction increased).

(iii) Tree Diagram (Work Breakdown Structure).

This is to generate alternative courses of options. When the alternative at each stage branches out, it becomes a tree-like diagram. When planning a project, a simple way of discover-

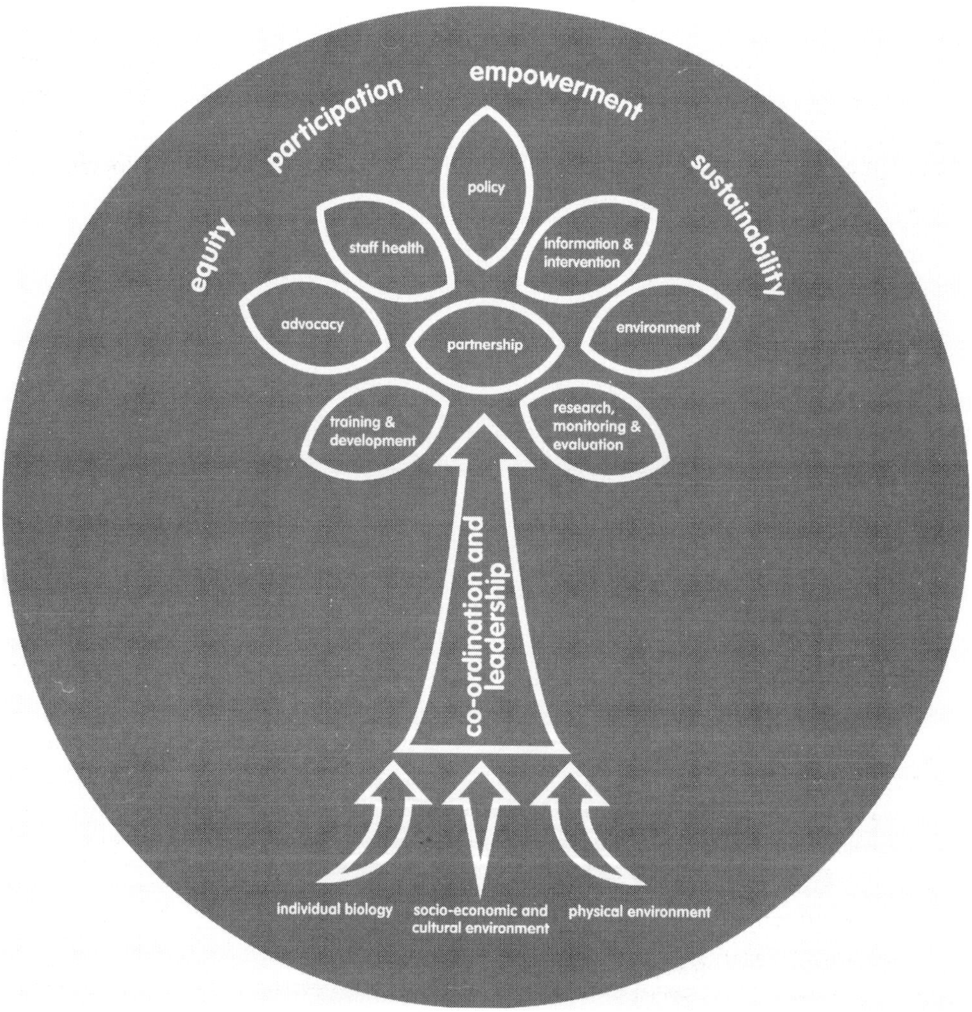

FIGURE 12.3c: Tree diagram.

ing the tasks to be performed by starting at the higher level and then breaking things down one level at a time. For example, a software project will start with requirements, design, etc. The design is of the architecture, verification, etc. The architecture is of user-interface, database manager, etc., and so on. This hierarchy can be shown in a Work Breakdown Structure (WBS). When you have got down to allocable units, you can then assign people and resources to each task and link these tasks together in an activity network.

(iv) Matrix Diagram – To clarify the relations between two different sets of factors. The matrix diagram is often used in deploying usage quality requirements into counterpart engineering characteristics.

The matrix diagram allows a many-to-many comparison of two lists, by turning the second list on its side to form a matrix. The illustration below shows how the relationship between two items can now be indicated in the cell where the row and column of the two items cross. Common extensions to the matrix diagram include the use of different symbols to indicate different comparison levels and the weighting of the items being compared.

There are a number of different shapes of matrix for comparing more than the basic two lists. These include the L-Matrix (the basic matrix as below), C-Matrix, T-Matrix, X-Matrix and Y-Matrix. The Matrix Diagram is a core tool in Quality Function Deployment (QFD). The matrix diagram is one of the new seven QC tools.

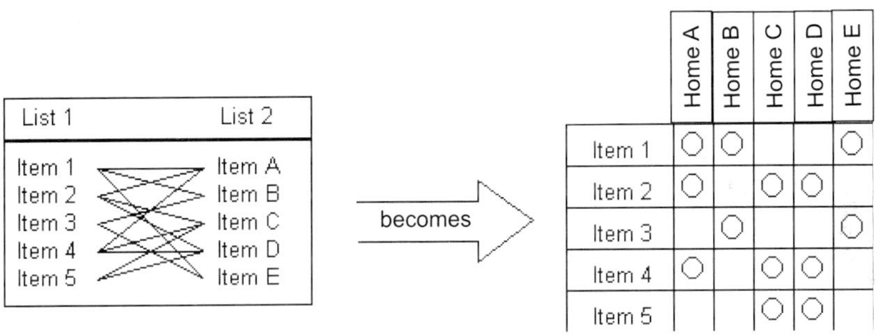

FIGURE 12.3d: Matrix diagram.

(v) Matrix Data Analysis Chart (MDAC) – This is an expanded version of matrix diagram, giving comparative weightage among different factors. By giving suitable weightage, the right mix of features can be incorporated into the product.

When comparing a large set of items, the complexity of the situation can make it difficult to determine how different factors relate to one another. In particular, it can be useful to find groups of items that behave in similar ways. For example, a washing powder may have different efficiencies at achieving 'softness' and 'stain removal' in garments made of acrylic, polyester, wool and various fibre mixtures. If similar affects are found in a group of fibres, then changing the powder ingredients may affect the whole group in a similar way.

The Matrix Data Analysis Chart (or MDAC) helps classify items by identifying two major characteristics common to all items and then plotting each item as a point on a standard x-y chart. This makes it easier to see how the individual items relate both to the characteristics and to one another, thus:

Measured Item	Character-istic A	Character-istic B
Item 1	10	8
Item 2	5	-4
Item 3	-8	-5
Item 4	-5	3
Item 5	7	-5
Item 6	8	9

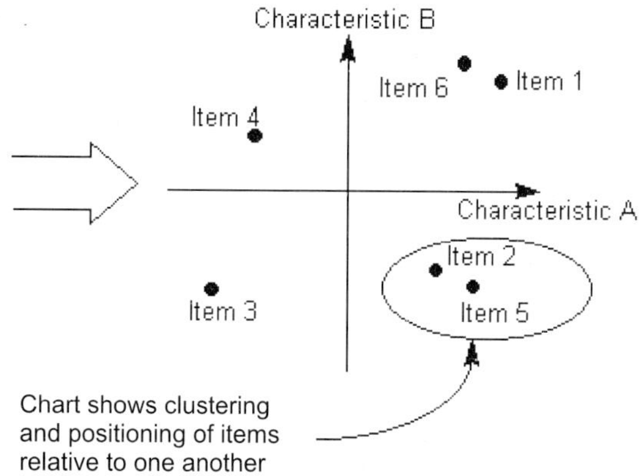

FIGURE 12.3e: Matrix data analysis chart.

(vi) PDPC – Process Decision Programme Chart. This is a tool for anticipating and dealing with unknown factors. This tool helps avoid surprises. The Process Decision Programme Chart (commonly abbreviated to the less unwieldy PDPC) provides a simple method to help in identifying both risks and countermeasures. Quite simply, if the plan is displayed diagrammatically (typically in a tree diagram), then identified risks and countermeasures are added in subsequent boxes, as below. This may appear to be quite trivial, but as with many other tools, the construction and thought involved is as important as the final diagram. The PDPC is one of the New seven QC tools.

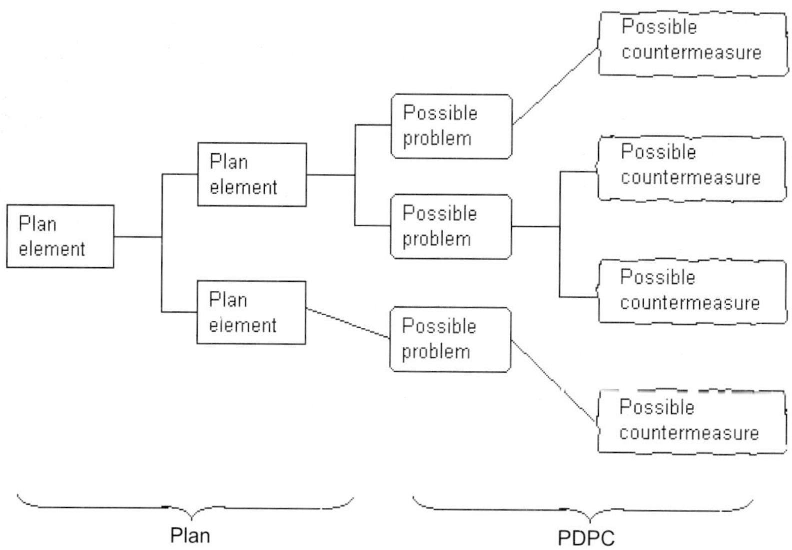

FIGURE 12.3f: PDPC diagram.

(vii) Arrow Diagram – For implementation schedules to be drawn up. It shows the necessary steps in a network form. One of the 'second seven tools', this is a classic project planning tool for piecing together a multi-task complex set of interdependent activities. It allows you to calculate what task starts when, what float there is in the project, etc. Also called an 'Activity-on-Node Diagram', the alternative (but potentially more troublesome) method of doing this is with an 'Activity-on-Arrow Diagram', or simply 'Arrow Diagram'.

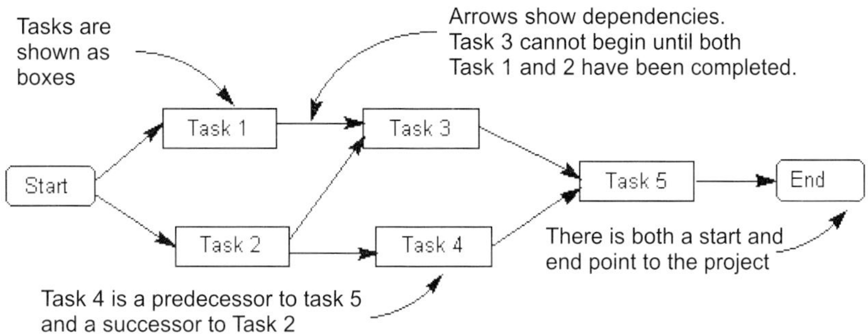

FIGURE 12.3g: Arrow diagram.

For TQM, these new tools are useful depending upon the areas of application.

The third pillar of TQM is employee involvement. Ultimately quality is physically produced by the operator on the shop floor, it is therefore important that he understands the quality requirements of the job. Instructions to him have to be clear. Without his cooperation, quality cannot be the outcome. It is here that the Japanese have introduced their Quality Circles concept which if implemented properly will generate a lot of commitment among workers.

In Indian organizations, there is a recent awakening on various approaches to this problem. Whatever may be the approach, the aim is clear, namely to have employee involvement. This may take the form of Suggestions Schemes, Quality Teams, Task Forces or whatever may be appropriate. Explaining in clear and simple language to the operators the significance of quality with reference to their particular job is more important than general lectures. For example, in a recent case where an organization had explained to its workers what the rejected material actually costs, the workers were taken aback and expressed that had they known this earlier, they might have taken better precautions!

Thus, TQM calls for a systematic and thorough approach to the three main aspects.

(i) Quality of service to the internal customers
(ii) Problem-solving tools to be applied so that facts are dealt with instead of opinions
(iii) Employee involvement.

12.2 AGILITY IN TECHNOLOGY

With an increase in the need for agile manufacturing along with the factors of short lead time, short product lives, increasing consumer awareness, etc. regarding the quality of the product, it is becoming increasingly important for the manufacturers to initiate steps to achieve agility in technology, especially in design, "Design anywhere, manufacture anywhere and market anywhere in the world" is becoming increasingly imperative for all competitive enterprises on the global challenge. The product design and manufacturing (consisting of **CE, RPT, Six-Sigma, SMED, JIT,** etc.) and Management Practices (such as TQM, TPM, BPR, etc.) are equally important and responsible for the products and services in this new millennium.

12.2.1 Agile Design

The dictionary meaning of the word 'Agile' is "Quick/Nimble/Active". We would however like to understand agile as an acronym of five words (viz.)

- Active (Energetic/Diligent/Originating action)
- Global (in outlook)/Worldwide ["Global village"]/"Plan globally act locally"
- Intelligent Quickness in understanding/Deciding between right & wrong to figure out a new what-we-are & what-we're-about and decide whether our former core competencies are still worth polishing: Presupposes continuous learning from all directions
- Lean "Do more with less"/Increase output ever increasingly with ever decreasing inputs over time, such as Yearly Turnover/Employee; Yearly value added/ Employee
- Effective (We know that efficiency is doing things right. And effectiveness is doing the right things. Effectiveness therefore is more important than efficiency. A prerequisite for effectiveness is sharpening our saw on a continuing basis. In short we need to follow the **Olympic Motto (Faster & faster; Higher & higher; Longer & longer)**

12.2.2 Model of Agility

All factors must be in good alignment to be internally consistent and mutually supportive, then the organization is poised & energized for maximum agility in design and manufacturing. The design of products and design of manufacturing processes are a matter of life and death to any manufacturing organization. **Charles Darwin's Theory of "Survival of the fittest"** is equally applicable to the species of organizations competing for the favour of the customers in the globalized economy. Hence, there is the need for ever-increasing agility.

In agile manufacturing of a product, it is mandatory to use DFM and DFA concepts. The integration of CAD-DFM-DFA is an effective approach to :

(i) reduce the number of components in a product
(ii) reduce labour time
(iii) minimize the total cost of the product

DFA means design of the product for ease of assembly. Designing for assembly encourages designers to concentrate on ease of assembly, while designing the product. DFA is a systematic procedure to minimize the use of components in the design of a product. DFA can be applied by utilizing simple patterns of movement and minimizing the axes of assembly. Complex orientation and assembly movements in various directions should be avoided. Part features should be provided such as chamfers and tapers. The product's design should enable assembly to begin with a base component with a large relative mass and a low centre of gravity upon which other parts are added. Assembly should proceed vertically with other parts added on top and positioned with the aid of gravity. This will minimize the need to reorient the assembly and reduce the need for temporary fastening and more complex fixturing. A product that is easy to assemble manually will be easily assembled with automation. Assembly that is automated will be more uniform, more reliable and of a higher quality.

DFMA is a process used for designing products and the relating realization processes to optimize the relationship between design function, manufacturability and ease of assembly and in turn will help to reduce the manufacturing cycle time and ultimately the cost to manufacture. The aim of DFMA is to maximize the use of manufacturing processes and minimize the number of components in an assembly or product. Principal benefits have been the reduction of development cost and increase in product quality and competitiveness. DFMA is a systematic procedure for analyzing proposed designs from the perspective of assembly processes. To obtain the maximum benefit from DFMA, the procedure is applied as early as possible in the design process and used within a CE teamwork environment. In conjunction with the procedure, designers can make use of DFMA guidelines to help manage and reduce the large amount of information involved.

Agility in technology needs extensive computation and Information Technology such as:

(i) Computer Aided Design (CAD)
(ii) Computer Aided Process Planning (CAPP)
(iii) Computer Aided Manufacturing (CAM)
(iv) Computer Aided Process Planning and Control (CAP), etc.

(i) Computer Aided Design (CAD)

The product cycle in an industry starts with the design. In the design process, the concept is modelled into the computer with suitable tools available for the purpose. One of the principal requirements of CAD is therefore the graphical representation of information. Graphical representation is always convenient from the standpoint of understanding. This applies to the design stage, analysis stage, simultaneously to the manufacturing stage. It is also necessary that the information entered once into the system should be made use of by all the other related activities. The various functions that are expected of the data generated during geometric modelling are:

Design Analysis

- Evaluation of areas and volumes
- Evaluation of mass and inertia properties
- Interface checking in assemblies
- Analysis of tolerance build-up in assemblies
- Analysis of kinematics – mechanics, robotics, simulation
- Automatic mesh generation for finite element analysis

(ii) Computer Aided Process Planning (CAPP)

CAPP is a means to automatically develop the process plan from the geometric image of the component. The key to development of such **CAPP** systems is to structure the data concerning part design, manufacturing facilities and capabilities into categories and logical relationships. **CAPP** thus integrates **CAD and CAM.**

(iii) Computer Aided Manufacturing (CAM)

The steps involved in generating manufacturing information from a CAD data model are required for defining the machining path through the intervention of the user where by the precise CLDATA (Centre Line of the Cutter Data) to be passed on to the postprocessor for generating the part program.

(iv) Computer Aided Production Planning and Control (CAP)

The information needs of the manufacturing system, in particular for a small to medium batch manufacturing system with a large variety of discrete components would be enormous. The conventional methods of handling such data invariably causes delays in identifying the problems as well as suggestions for providing the remedy to the situation. The solution in such cases is the use of computers which are now available at sufficiently low cost and have enough required processing capacity. The adoption of information management system do not incur a large outlay of funds as is the use of advanced manufacturing techniques such as CNC machine tools and robotics.

Though computerization is attempted in a number of medium- and large-scale units in India, it is always done piecemeal, e.g., financial management function such as costing, purchasing, personnel, etc., Not much effort is evident in the engineering aspects (factory management). By such isolated computerization, the overall benefits that could be achieved by proper integration of the complete information system could not be derived. The complete integration essentially calls for a proper discipline and streamlining of the information handling in the factory. The major problems that are faced by manufacturing industries due to lack of proper information integration are:

- Confusion and congestion in shop floor
- Large throughput time
- Excessive and unnecessary handling
- Large work in process inventory
- Unnecessary cost addition
- Poor utilization of resources

Complete benefits cannot be achieved by developing isolated computerized systems. This calls for a need to develop a complete integrated information management system. The information integrated system not only optimizes the performance parameters, but also provides the necessary information at the right time for taking the decisions.

Case Studies from Literature

(i) A. Bhatia and P.V.M. Rao (2002) carried out a case study on a timepiece and showed significant reduction in assembling time for the product.

(ii) Prof. M.C. Jagath and Mr. Gaurav Batra conducted a case study on an existing TV Remote Control and showed:

(a) Reduction of number of parts from 6 to 4
(b) Reduction in labour time from 60 seconds to 27 seconds
(c) Reduction in total cost per product from Rs. 36 to Rs. 23.

(v) Example of Agile Design

Coltec Industries, a Sallisaw, Okla, maker of engine components, needed a system to machine aluminium oil-pump bodies. The company required an agile system that could accommodate design changes and expand. Also important were simple maintenance, rapid response, and quick changeovers.

In response, Cargill Detroit Corp., Clawson, Mich., constructed a system consisting of multiaxis CNC machining stations and a free-flow parts transfer scheme using special pallet fixtures. "The quality of the parts and the flexibility and uptime of the system are far in excess of what we had achieved with conventional machining," says Pete Johnson, Coltec Project Manager.

12.2.3 Competitive Advantage through Design

As the process of globalization is taking momentum, the demand for design inputs from varied sectors has brought in the new concept of design economy leading to better awareness of well-designed quality products and augmented services. Consumer preferences, lifestyle changes, economic liberalization and technological advancements are bringing in new opportunities for designed companies. Design should be capable of integrating current technological inputs and innovative solutions concerned with the appropriate use of raw materials, increased productivity, improved communication, health and human safety, improved functionality preservation of natural and cultural resources, enhancement of work and living environments. It has a place especially when work opportunities and earnings are expanding at every level of the economic activity.

The companies which are 'innovative' and have focus on 'design' and 'technology' to achieve competitiveness and those which give importance to perceived needs of the customers will survive in the new economy. The design defined as generation of knowledge plus application of new knowledge which encompasses all functional areas ranging from design to manufacture, to products, to marketing, to management and technology development, innovation is the engine of growth for any organization or company. Here it is the design which meets the perceived needs through creation of technical solutions to problems.

Throughout the world, good product design is fast becoming recognized as a basis for securing competitive advantage for manufacturing industries. In recent years, the subject of design and product development has been the focus of companies. The government in several developed and fast developing countries aimed specifically at bringing about improvement in business performance. For many years, organizations tended to concentrate on improving the functional performance of their manufacturing processes. But, in the new millennium, radical improvements can be brought about only through better organization, management and control of the design and new product development processes.

12.2.4 Emergence of Design for Competitiveness

Today, in most products, the technological developments are mainly horizontal rather than vertical. This has led to a situation where different companies vie with each other in quality, performance, reliability and price. In such a situation, where all the other parameters being equal. Industrial design can provide the necessary product differentiation and perception for the company to effectively compete in the market. Design thus plays a crucial role in making early inroads for the market success of any business.

However, design has a lot to do with products

- Performance
- Quality
- Durability
- Manufacturability
- Serviceability
- Appearance and
- Cost

The word 'design' is seen to be directly associated with consumer products and fashion and thereby with the visual appeal or aesthetics of the product. But, design is more than skin deep or a cosmetic activity which the manufacturer will use in the last phase when everything is finalized and the product is about to be launched. Design is at the very heart of a successful product. A good design should not only appeal to the eye, but it must also make the product reliable and efficient. For the manufacturer, it should also be easy to produce, assemble, use and maintain.

Increasing specialization has forced engineers to concentrate on the engineering and technical aspects so as to say the 'functional aspects' of the product, while industrial designers take care of the aspects which have direct contact with users, ensuring that the product satisfies the users.

Industrial design is a profession that specializes in optimizing the function, value and appearance of the product for the mutual benefit of the user and the manufacturer. Industrial designer thus looks at the product from the user perspective and concentrates on the usability of the product.

What makes a product "whole", as perceived by the customer is its so called "physical and reliable performance", "user-friendly design" and "better mental feeling" that it brings to a user.

Design has some indirect effect on indicators of international competition, level of consumption, research and development, innovation, commercial balance, technology and also collective well-being. Design develops a demand for consumer goods that stimulates national consumption and international competition.

Technology, globalization, communication convergence and consumer empowerment are some of the challenges confronting India's manufacturing sector, compelling organizations to examine new strategic formulae. In their search for answers, design has surfaced as the foremost strategic tool, combining corporate objectives with consumer satisfaction. Design forms an integral part of a market leader's business strategy. It offers a substantial value enhancement that brings about widespread consumer acceptance, creates markets and generates brand loyalty.

In the international arena, competitiveness requires adaptation to a multicultural context with strong regional structures for reviving and encouraging different ways of life to enable a network to survive in the global trade.

12.2.5 Better Design for Manufacturability/Assembly/Serviceability

Although the cost of design is perhaps the lowest portion of the total product cost, said to be about 8%, a major part of the product cost, said to be about 80%, is committed at the design stage itself. The decisions arrived at this stage greatly influence the ease or hardship with which subsequent activities of product development and manufacture are carried out. Apart from these activities, the design decisions also influence other activities during the life of a product up to the stage of recycling. Reorienting the design process so as to make activities like Manufacture, Assembly, Mass Customization, Safety, Serviceability and Environment make the product development and life cycle more effective. The designer also needs to be involved with other functional departments such as marketing and engineering services through CE and collaborative engineering methods.

What is required is a knowledge-based approach that attempts to design products that maximize all desirable characteristics — such as High Quality, Reliability, Serviceability, Safety, User Friendliness, Environment Friendliness and Short-time-to-market in a product design while at the same time minimizing manufacturing costs and of course, lifetime costs. Designers today have to keep an eye on the downstream considerations that affect usability and cost and stop emphasizing only on functions, features and appearance. Timeliness, quality and affordability are the three elements of composition of competitiveness which also determines the value of the product for the consumer. Design needs to be continually reviewed from the start to the end to find ways to improve production and other non-functional aspects.

The advantages of these techniques are shorter production times, fewer production steps, smaller parts inventory, more standardized parts, simpler designs that are more likely to be robust and that can work even without expertise. The salient point here is that in order to capture the competitive advantage, the designer has to meet many characteristics and parameters.

12.2.6 Information Technology (IT) for Design

The Internet, it has been said, is a stunning combination of knowledge aggregation and infrastructure de-segregation. More and more information is available to more people, more quickly and with less effort. There is more data than any one person can grasp, manage, make sense to designers, like other professionals, have been grappling with this phenomenon in various ways. Digital design will play a major role in the transformation of the world's economic landscape. Digital design is just a new way of performing the design function. In order to respond to competition and change, digital design is seen as a critical capability to be developed especially in developing countries. We need high quality design capabilities which need to be adequately strengthened. The IT sector is at a critical stage as a user of design when it comes to the creation of new and improved products at the upper end of the value chain. The pressure for trained designers in this sector is increasing as industry starts addressing the higher value segments of the market.

There are two ways digital design is going to progress, one is through the application of digital tools and processes for the traditional economic sectors and the second will represent the use of design for the creation of new information tools and services for the IT sector itself. The discipline of design will require high quality original technology to produce excellent product with innovation. Design is the new medium which is being transformed by the use of digital technologies and design educators must be encouraged and supported to develop new and improved processes through an intensive programme of investment in design education and research. Designers need to equip themselves with the required skill to work in this mode of design action.

12.2.7 Creating and Enabling Environment for Design and Innovation

Innovation calls for an intimate understanding of current customers and markets and potential new customers and markets. It also includes team and organizational competencies and improvement opportunities, visions, values and missions. Expenditure on research and development is a key indicator to the level of economic activity in an economy. This expenditure is directly linked to the quantum of earnings and gross margins are linked to past expenditure on Research and Development. It is not only the amount spent on Research and Development which is important, but also the effectiveness of the spend on the new products that result.

In today's competitive world, increasingly leading industrialists and chief executives are recognizing the importance of technology management to the future of their business, not technology for its own sake, but technology that can be converted into innovative product ideas which can in turn become attractive, well developed and reliable satisfier of customer needs.

Effective innovation depends on disciplined management systems and processes, that encourage a search for creative ways to do things better, different or more effectively. Beyond the management tools of surveys, focus groups and innovation leaders find a multitude of ways to understand and satisfy their customers. So many new products and services come from emphatic innovation which flows from a deep empathy and understanding of the intended customers' problems and aspirations.

The control of technology through its conversion to top class new products through a high level of development competence is crucial. By paying attention to the process of innovation rather than just the product itself, the design and development team can move from fire fighting to controlled management.

The need for rapid and effective new product development is driven by increases in aggressive global competition; the development and application of new technologies; shorter product life cycles, increasing customer choice and customization; and more technology inputs.

12.2.8 Designing for Global Opportunities: Corporate Perspective

In the process of liberalization, India is moving at a faster pace from a protected market to a market where preferences are available to users more than ever before. Any equipment, therefore, has to "Emphasize in the Crowd" to get a user to buy. Industrial design can contribute to a greater extent to make this possible.

With technology transfer by multinationals in India, a suitability of a particular design for use in the Indian context is not always considered. As an outcome of this situation, equipment developed abroad invariably had to be specially adapted to meet the ergonomically and psychological needs of the Indian user. This can now be achieved as there are trained industrial designers in the field.

In the Indian context, there is a greater need for diversification. Design can essentially work towards this goal. Industrial design can enable industry to effectively face the competition in the marketplace. The industrial designer has to be given an effective role in equipment planning and development so that the manufacturer can satisfy the end user.

Though a decision is taken in the in-house studio, the services of a design consultant are still necessary, since the interaction of an outside consultant with in-house design team of an industry gives direction to the design approach.

A company's commitment to design is not restricted to equipment done. It is also reflected in several other important aspects: commitment to customers and society through constant effort on well-designed products and services, its building, the workplace organization, sponsorship of design awards, and its overall visual image. All these put together must make a statement which spells quality, excellence and innovation. This is corporate design in the broadest sense.

12.2.9 Knowledge-Based Design (KBD)

Knowledge is not simply the accumulation of facts, information or data, but the application of them. Thus, KBD is an automated system that provides for the capture of process specific knowledge gained throughout the product's life cycle and applying that knowledge to product specific decisions. KBD overcomes the problems and barriers presented by CAD in automating the application of this accumulated knowledge.

- Increased technology utilization has resulted in decreased communication.
- Decreased communication negatively impacts performance, quality and cost.
- KBD process is better than CAD process.
- You can and should make the paradigm shift to KBD.

- **Increased Technology, Decreased Communication**

  It is interesting when we look at the evolution of "Computer Aided" technology that the first process to be automated was manufacturing (CAM). Innovators found that smaller than mainframe computers could be used for numerical control of simple machine processes on the factory floor. With the advent of graphic terminals in the late 1960s and early 1970s CAD entered the scene. This is largely due to the fact that the computers were not very powerful and drafting was not a very mathematically challenging task. When sales of CAD came slowly, it was a brilliant marketing move on the vendors to rename CAD to Computer Aided Design. The market has not changed, only the market perception of what the technology could do. With this introduction of CAD, the world of engineering began a journey of increasing technology and decreasing communications.

- **Typical Design Process**

  Let's begin with a look at the typical design process in most companies today. When an engineer gets an idea, he quickly sketches the idea on the closest article he can carry and then passes it on to a designer to make the pretty pictures. These pictures, in the form of CAD data bases, are then given back to the engineer for review, corrections are made, changes are incorporated, etc. This process continues for a given period of time.

Once the design is deemed "build-table", it is turned over to manufacturing. Manufacturing then has to convert the drawing to a format they can understand, re-establish data, etc. At this point, they may realize that they can't build the design. This information is given back to the engineer with the appropriately censored notes, which then spawns yet another cycle. Once this communication is successfully completed, the design is then sent back to the designer for changes.

Now we have a three-way loop of moving from engineer to designer to manufacturing, to engineer, etc., which can continue for an indefinite period of time. If you follow this loop closely enough as shown in Fig. 12.4, you will see a suspicious likeness to the symbol for infinity.

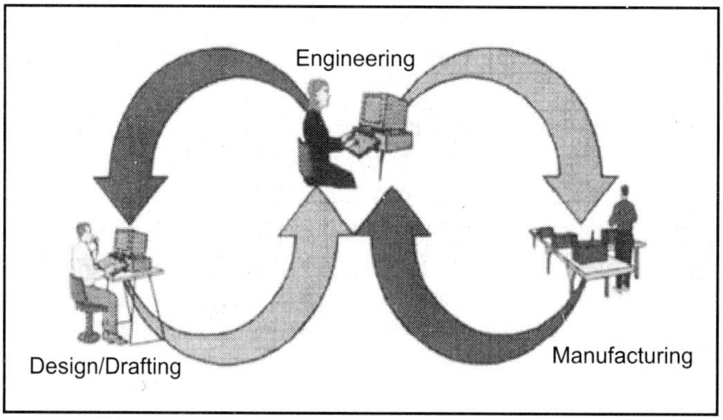

FIGURE 12.4 Typical design flow

Finally, someone makes the decision that the product must be delivered at which point everything is finalized, built, and shipped. Is it any wonder why we have developed an attitude of never buying a first version of a product.

- **Islands of Automation**

Automation began to accelerate for engineering and manufacturing companies. With the advent of cheaper and more powerful computers, technology began to emerge to automate these "tasks". As they get matured and industry accepted these technologies, we created more and better ways of doing the same things. The end result is highly refined process steps in the life cycle of our products with little or no improvement in communicating that new found knowledge. We have created very sophisticated "Islands of Automation". An engineer can now create geometry much more quickly and easily than ever before. A draftsman can create accurate drawings at a virtual push of a button. Manufacturing can create parts automatically using the advanced CAM technologies. However, what has been missed is capturing and sharing the "lessons learned" during the process.

- **Concurrent Engineering (CE)**

To address this issue, we created "CE Teams", "Integrated Process and Product Teams", and any number of other descriptions we could cleverly come up with, to describe capturing and sharing of our lessons learnt. These teams usually comprise representation from each necessary discipline (i.e., Design Engineering, Manufacturing, Quality Control, etc.). The overriding feeling is that this brings the collective "knowledge" of the company to the project. This approach has tended to fail for two reasons.

- Firstly, we require each member of the team to possess all the knowledge necessary to address any and all questions of their specific departments when they come to the meetings. We expect the manufacturing person to know everything about every process in the shop and be able to address any questions that may arise. Knowing, this is very unlikely, they must then defer the answer until they have a chance to track down the appropriate specialist with the answer and then return to yet another meeting to respond. This is then repeated any number of times with each group represented.
- Secondly, we expect everyone in the team to get along and cooperate. We have created this false image of a group of people skipping hand-in-hand down the hallways knowing all, telling all, hearing all, and doing all – not likely.

- **Negative Impact of Decreased Communication**

Many companies have held to the premise that knowledge is power. This has become more of a political statement than a technical one. The end result of holding knowledge within your own organizations has helped create this chasm in sharing knowledge. The end result is less quality and reliability with greater costs.

- **Cultural Barriers**

Aside from the obvious logistics issues, there are two primary barriers that have kept the knowledge gained during downstream operations from being shared and implemented within the engineering community. Firstly, the technology to capture this knowledge and make it easily available to influence or drive decisions by other departments has not been readily available. Secondly, the adversarial cultures between engineering and manufacturing would now allow it as shown in Fig 12.5.

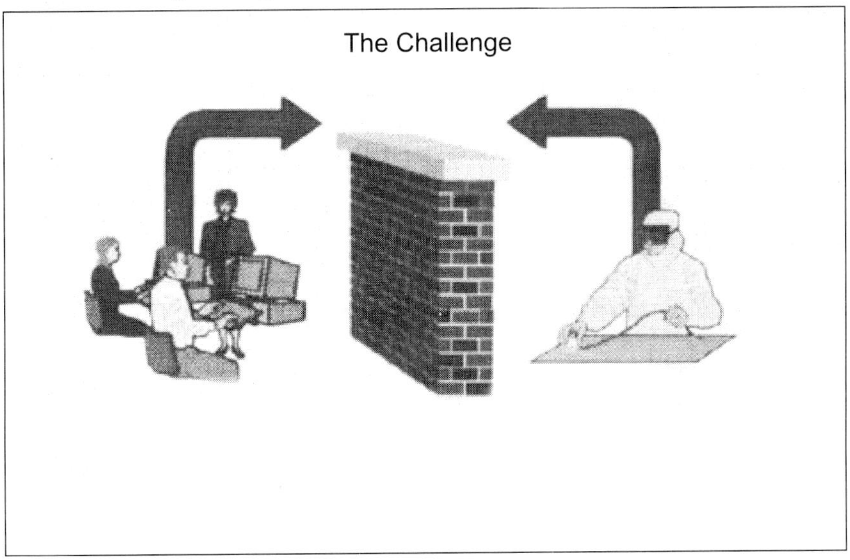

FIGURE 12.5: Adversarial cultures between engineering and manufacturing.

To highlight this barrier, *Business Week* recently reported that, "A high wall has always stood between design and manufacturing, with designers essentially ignorant of how their creations are transformed into products". Another article published in *Automation* magazine stated, "The typical American factory spends one quarter of its operating budget to fix mistakes. Yet only 20% can be traced to the way we produce our products. The other 80% rests largely in the way we design them".

- **Performance, Quality and Cost**

As we go through the design, manufacture, delivery, and support of products we gain knowledge. By the time a product moves from the conceptual stage to the detailed design stage, 80% or more of the cost of that product and most of its quality, reliability and manufacturability have already been determined. The challenge is how to make that knowledge gained at the back-end of this process available to the designer as he/she is making critical design decisions.

Not only do we gain knowledge during the life cycle of a product, but as shown in Fig. 12.6, the longer we go in the process, our decisions have less impact on the product's performance, produceability, costs, reliability, quality, and maintainability. This lack of impact directly affects the corporation's "bottom line".

Most of the changes which directly impact the products performance, quality, etc., which are learned along the way, are not implemented due to the cost of making those changes. These changes are not ignored due to any lack of desire to produce quality products, but a reduced window of opportunity due to market demands, customer expectations and competition. Therefore, we must take advantage of this brief window of opportunity by implementing knowledge based tools allowing us to make design decisions early in the product's life cycle where they will have the greatest impact on quality, reliability and produceability while minimizing costs.

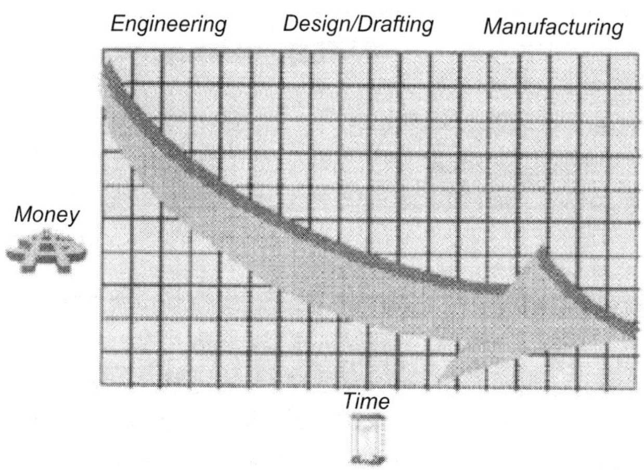

FIGURE 12.6: Impact of design decisions

- **KBD is Better than CAD**

What has "Computer Aided Everything" really done for us? To understand this question, we must first look at pre-CAD engineering approach to design.

- **The Pre-CAD Engineering Process**

A traditional design process would do something like this. Someone in sales takes an order for a product we don't have. The engineer, using this functional specification, makes a basic sketch. The sketch is then taken to the shop where manufacturing makes a crude prototype of the design. The prototype is then put through its paces. Knowledge is gained at each step of this process. That knowledge is then used to make changes to the form of the design. This cycle continues until a working prototype is accomplished. Once a working prototype exists, a draftsman is sent to the shop to measure each piece and document a successful design. This methodology puts the focus of the design on its function rather than just its form. Drafting or documentation of the design was the last step in this process.

Traditional engineering practices taught in our universities have been that an engineer begins a design project with two pieces of information;

(i) Performance criteria - what does this product have to do? and;

(ii) Design rules: what are the physical constraints?

By using these two pieces of information the engineer must then come up with a design that does what it was supposed to do and fit into the physical environment required. The form of the design is secondary at this point.

- **What does CAD bring?**

The CAD focus, however, has been a geometry (form) based design process. The focus has tended to be on the technologies that create geometry and not on the product being developed. The issue should be on improving our documentation of a nine-element mechanism, but finding a way to achieve equal or superior performance with a five-or-seven element solution, so that significant cost reduction can be achieved while increasing the reliability and quality of the product.

When the CAD approach is used, we often fall into the "Build-Test-Break" cycle (or what I like to call the "Grease and Grind" mode) of product development.

By placing Form first, the activities related to function (synthesis, analysis, and testing) are eliminated from the process. The three "F's" of design Form, Fit, and Function, get reduced to Form and occasionally Fit and that is only if 3D CAD tools are properly used.

The "new" introduction of old technologies, namely, Solid Modelling and FEA, were still a geometry-based approach to the design process. This is what we have depicted in Fig. 12.7 as **"Form over Function".**

Again, CAD allows us to easily determine the form and sometimes fit of our designs but still does little for the function of the design. By determining the form of the design ahead of its function, engineers are limited to changes that can be made in their design. The process is then relegated to utilizing a solid model as a soft prototype, analyzing for strength, weight, etc. and then making whatever changes are needed to make the design work. The chances of a fully successful design are limited by the initial decisions about the form of the design. As depicted in figure earlier, an infinite cycle begins that usually results in products being released and delivered before they are fully functional.

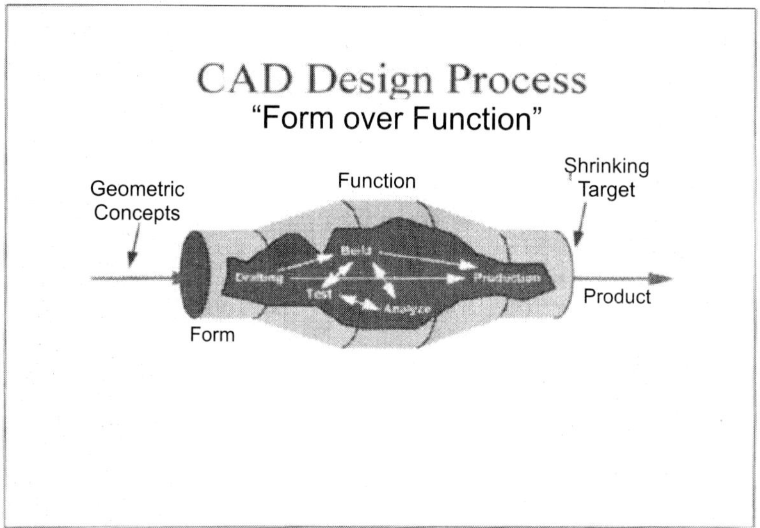

FIGURE 12.7: Design process using CAD

- **What does KBD bring?**

KBD is an automated return to traditional engineering practices when approaching product design. This traditional approach is to consider "Function over Form" as shown in Fig. 12.8 when making design decisions. What good does it do to have a beautifully sculpted design which does not function as advertised.

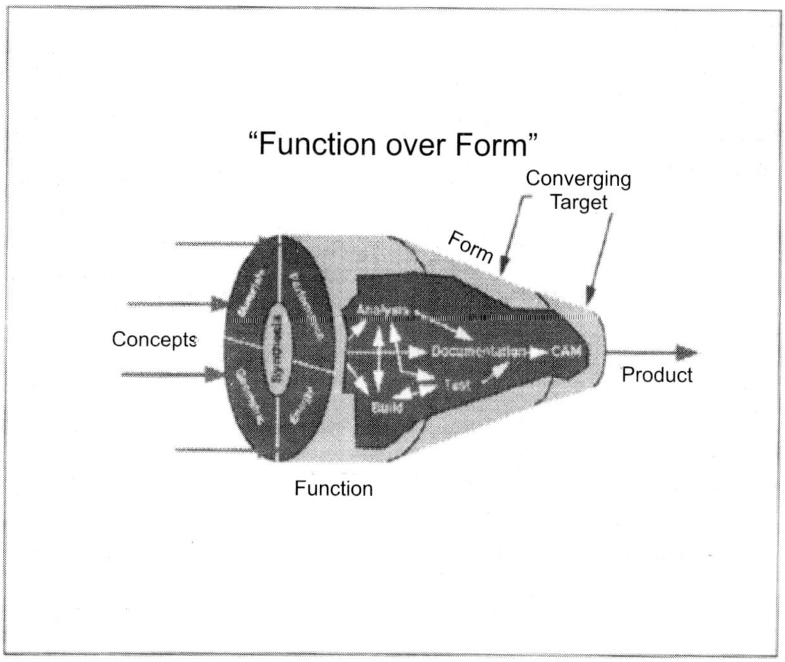

FIGURE 12.8 KBD design process

KBD provides for complete synthesis of the design, i.e. a computer-based working prototype to verify and optimize the performance of the design while being kept fully cognizant of the associated costs and issues pertaining to producibility of the design. One of the keys to successful product innovation is to rapidly achieve a design conclusion.

The real opportunity of KBD is to correct manufacturing problems (where they are often caused); at the design stage itself and not in the factory (see Fig. 12.7). If engineers have tools to better make use of their time to analyze the performance of a design, to investigate potential manufacturing problems or to concentrate on improved quality, the result would be better design and better products.

By performing these studies from an integrated knowledge base rather than typical trial and error, grease and grind, build and break, engineers can receive instant instruction as to best practices and methods based on lessons learned.

- **Who has the knowledge?**

YOU DO! You already possess all the knowledge necessary to make your products better, faster and cheaper. The overriding need is not new and innovative ways to perform specific tasks, but automating the sharing of the knowledge we already have in-house and have that knowledge affect our designs while the critical decisions are being made, upfront.

We have filled rooms with volumes of information about engineering practices, drafting standards, material properties, mathematical formulas and equations, but with little assurance of that information being used. You can also walk into almost any manufacturing environment and instantly get more suggestions than you would ever want to hear about how to improve the design process. As you move on to test and validation and then to the support organizations that have to make the design work after it gets to the customer, you'll gain even more insight.

- **What about cost?**

Let's not forget about cost. Due to competition, today's market has changed significantly. No longer can we rely on traditional cost plus models that states, "How much does it Cost us to build?.... What is our target Profit Margin? .... Add these to get our Price", or....

Cost + Profit Margin = Price

We are now forced to consider cost as an independent variable in our designs. Today's formula looks more like this, "How much can we sell it for? .... What is our target Profit Margin? .... Subtract these to get our target Cost", or....

Price – Profit Margin = Cost

Design to Cost (DTC) is a reality, with little or no understanding of how to address the need.

- **You Can and Should Make the Paradigm Shift to KBD**

So, what is stopping you?

All you really need to do is to capture all the knowledge contained in your engineering manuals, combine that with the real-world experiences of your engineering staff, add to that the combined knowledge of all your process specialists, industrial planners, schedulers and manufacturing engineers, include the lessons learned from your test and validation group, along with information from the support organization, and simply make that available on the engineers desktop at the push of a button shown in Fig. 12.9.

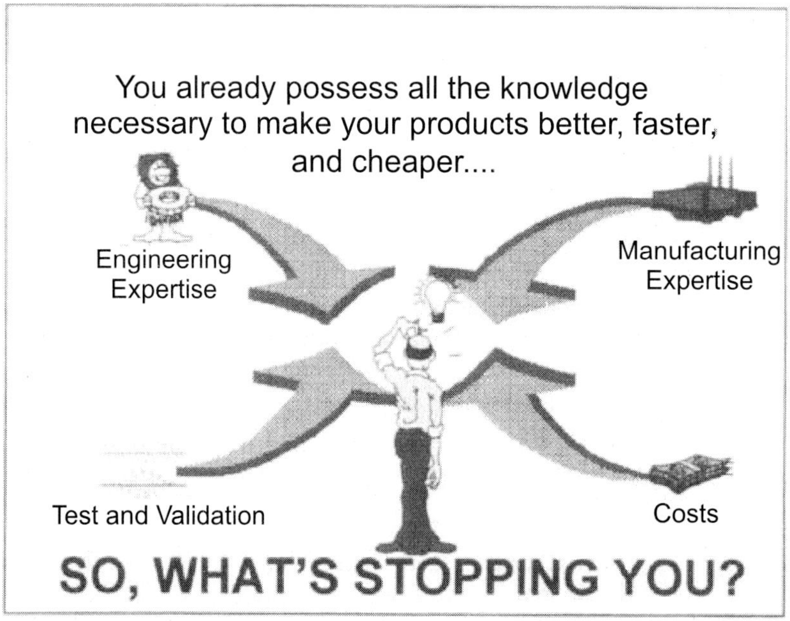

FIGURE 12.9: Paradigm shift to KBD (automating the process).

You cannot have true CE until you have automated this process. CE in most companies has become simply another term for meetings.

- **Is there an answer?**

Yes! There are tools and methods which exist for capturing, applying and automating your knowledge.

Cognition Corporation offers a complete **"Knowledge Centre"** environment for the design and manufacturing arena **(KC-2000)**. The KC-2000 is an expert object oriented system allowing the capture, manipulation, distribution and use of knowledge within a corporation. Keep in mind, that no computer-based system will ever completely replace the innovation and creative abilities of the human mind. However, the need to capture and use existing innovations is paramount.

The KC-2000 comprises three primary entities; Mechanical Advantage, Cost Advantage, and Work Advantage.

(i) **Mechanical Advantage (MA)** is a mechanical KBD system. It supports an engineer in formulating the design specifications for a product, creating various design alternatives early in the design process, fully analyze them, optimize on a chosen alternative and then fully evaluate the optimal design solution. It allows the engineer to develop a fully functioning "soft prototype" from which to evaluate the performance and behaviour of their design. During later stages of product development, the MA assists engineers in evaluating design details or in responding to failures in field operation or in engineering testing.

MA provides a design environment in which any length, angle, dimension or functional parameter can be an independent variable, the engineer has the ability to drive his design through equations, external programs, table data or graphs of sample data. The engineer has the ability to step variables through ranges, optimize variables for specific values, perturb a variable around a known solution for sensitivity analysis and to create design animations showing how a design would work or how the values of dependent variables change.

You can think of the functional capabilities in three categories; Performance Modelling, Tolerance Management, and Critical Parameter Management.

- Performance Modeling includes such functions as Motion Studies and Synthesis, Automated Design Optimization, Cam Verification and Synthesis and Complete Force Analysis. By linking mathematical equations and rule sets directly to geometric models, an engineer cannot only analyze his/her design but allow the system to synthesize the most optimal design.
- Tolerance Management allows the engineer to consider the effects of "Dimensional and Geometric Parameters" on a design. He/She can also perform push button tolerance analysis simultaneously looking at Worse Case Max/Min, Root Mean Square (RMS) Probability from 3 to 6 Sigma, and Linear/Non-Linear **Monte Carlo Analysis**. They can also give detailed insight to the Cp/CpK rating of the design. These capabilities are a result of the knowledge captured in the design process.
- Critical Parameter Management allows the engineer to also consider the effects of "Functional Parameters" on the performance of the design such as... inertia, voltage, weight, pressures, etc.

(ii) **Cost Advantage (CA)** is a rule based expert system specifically geared toward the cost and produceability arenas. Cost advantage will provide automated cost assessment for both parts and assemblies, produceability analysis, design guidance, manufacturing alternatives, DFMEA review, easy identification of possible cost drivers, automated trade studies, and provide detailed summary reports and audit trail.

It allows the reader to easily capture your knowledge and expertise about your design and manufacturing processes and also to include the knowledge and expertise of your vendors for those processes which you may be outsourcing.

It can easily interface to existing database information which may exist, such as Materials Libraries, MRP systems, etc. Furthermore, through a feature we call CostLink, Cost Advantage will perform its analysis and provide real time feedback to the engineer directly from their CAD system environment. Today, Cost Advantage has been linked to MA, Pro-E SDRC Master Series, Mentor Graphics, Unigraphics, and CATIA software systems.

Due to the extensive development recently at cognition in the area of data mining with a variety of CAD systems, we have acquired significant understanding into the interactivity between Cost Advantage and the CAD arena. The inherent nature of Cost Advantage to easily capture rule sets, coupled with our understanding of the CAD environment has allowed us to extend these design rule sets to directly drive the CAD environment. We refer to this practice as "Direct Engineering".

Direct engineering allows the end-user to capture the rule sets, usually available via Company Design Handbooks into a CA model. The engineer then interacts with the "forms" screen within CA to specify the functional requirements of the design. Once the design passes the evaluation from the rule set, CA then directs the CAD system to generate the appropriate geometry (CAD model) and associated drawings. This is particularly effective for families of parts. To date, several Direct Engineering projects have been completed in conjunction with software packages.

(iii) Work Advantage (WA): The final piece to this Knowledge Centre puzzle is our **WA**. By being based on an object database, The WA provides the flexibility to send and retrieve objects or cause actions (methods) to operate on other vendors objects. This is accomplished while replacing custom interfaces and **API's** with a standard open protocol such as **CORBA**. The system allows the end user to quickly and easily create schema's without requiring recompilation or highly skilled programmers. Furthermore, the system is Internet linked, contains a native browser and has been proven to be scaleable for multiple users with no degradation to system responses.

An example of its use would be; data extracted from the CAD system via Cost Link such as features, characteristics and values which can be placed in the WA. From that neutral object environment, the data can then be queried, extracted and used by any other application in the organization, i.e., NC tool paths can be generated from mathematical data rather than requiring a direct CAD link.

- **Observations**

(a) Increased technology has resulted in decreased communication
- ➢ Typical design flow is an infinite loop process.
- ➢ Technology has created islands of automation.
- ➢ CE teams are the result of a change in technology without corresponding change in the design flow.

(b) Decreased communications negatively impact performance, quality, and cost
- ➢ Decreased communication results in decisions about performance, quality, and cost being made late in the design cycle.
- ➢ Decisions made late in the design cycle cost more and are less effective.
- ➢ Decisions made early have a positive effect on performance, quality, and cost.

(c) KBD process is better than CAD process
- ➢ CAD is Form over Function.
- ➢ KBD is Function over Form.
- ➢ Function over Form is better.

(d) You can and should make the paradigm shift to KBD
- ➢ You have the knowledge to move from CAD to KBD.
- ➢ Tools and methods exist for capturing, applying and automating that knowledge.
- ➢ Cognition offers an integrated environment for KBD (the Knowledge Centre).
- ➢ MA, CA, and WA are examples of a KBD system.

12.2.10 Axiomatic Design

Prof. Nam P. Suh, MIT, USA has offered his own strong endorsement of **Axiomatic Design Principles** for faster bug free manufacturing of hardware and software systems.

12.2.10.1 Definitions

Axioms are valid only within the bounds established by the definitions of the key terms. The definitions are:

- **Axiom:** Self-evident truth or fundamental truth for which there are no counter-examples or exceptions. An axiom cannot be derived from other laws or principles of nature.
- **Corollary:** Inference derived from axioms or from propositions that follow from axioms or from other propositions that have been proven.
- **Theorem:** A proposition that is not self-evident but that can be proven from accepted premises or axioms and so is established as a law or principle.
- **Functional Requirements (FRs):** FRs are a minimum set of independent requirements that completely characterize the functional needs of the product (software, organizations, systems, etc.) in the functional domain. By definition, each FR is independent of every other FR at the time the FRs are established.
- **Constraint (Cs):** Cs are bounds on acceptable solutions. There are two kinds of constraints: Input constraints and System constraints. Input constraints are imposed as part of the design specifications. System constraints are imposed by the system in which the design solution must function.
- **Design Parameters (DPs):** DPs are the key physical variables (or other equivalent terms in the case of software design) in the physical domain that characterize the design that satisfies the specified FRs.
- **Process Variables (PVs):** PVs are the key variables (or any other equivalent term, as in the case of software design, etc.) in the process domain that characterize the process that can generate the specified DPs.

12.2.10.2 Mapping from Customer Needs to Functional Requirements

Once we identify and define the perceived customer needs (or the attributes the customer is looking for in a product), these needs must be translated to FRs. This must be done within a "solution-neutral environment". That means that the FRs must be defined without ever thinking about something that has been already designed or what the design solution should be.

After the FRs are chosen, we map them into the physical domain to conceive a design with specific DPs that can satisfy the FRs. The mapping process is typically one-to-many process, that is, for a given FR, there can be many possible DPs. We must choose the right DP by making sure that other FRs are not affected by the chosen DP and that the FR can be satisfied within its design range.

During the mapping process involving a set of FRs going from the functional domain to the physical domain, we must make the right design decisions using the Independence Axiom. When several designs that satisfy the Independence Axiom are available, the Information Axiom can be used to select the best design. When only one FR is to be satisfied by having an acceptable DP, the Independence Axiom is always satisfied and the Information Axiom is the only axiom the one-FR design must satisfy.

The basic postulate of the axiomatic approach to design is that there are fundamental axioms that govern the design process. Two axioms were identified by examining the common elements that are always present in good designs.

(i) The first axiom is called the **Independence Axiom**. It states that the independence of FRs must always be maintained, where FRs are defined as the minimum set of independent requirements that characterize the design goals.

(ii) The second axiom is called the **Information Axioms**. It states that among those designs that satisfy the Independence Axiom, the design that has the smallest information content is the best design. Because the information content is defined in terms of probability, the second axiom also states that the design that has the highest probability of success is the best design. In an ideal design, the information content should be zero to satisfy the FR every time and all the time.

As many case studies show, the performance, robustness, reliability, and functionality of products, processes, software, systems, and organizations are significantly improved when these axioms are satisfied.

(Refer to "Theory of Complexity and Applications", by Prof. Nam P. Suh, MIT, USA)

12.3 AGILITY IN WORKFORCE

While most employers cling to a business-as-usual mentality, complaining of inadequate worker skills, the workers themselves have yet to respond purposefully. Surveys of working people underline their anxiety about job security, restricted access to training and an unclear connection between training and long-term prospects for earnings, promotions or keeping the jobs they have. Workers surveyed are ambivalent (mixed) about getting more training and about half are not willing or able to do much to obtain training. A few key findings illustrate this ambivalence:

- Most workers believe that their employers already provide more than enough training for them to do their jobs well and that their skills and education are underutilized.
- Significant numbers of workers feel that they would attend training only if their employers required it and only on company time.
- Many employees view "training" as negative, somehow connoting (suggesting) failure.
- Employees who have participated in some form of structured training over the years are positive & feel it was very useful and they want more.
- Formal training generates good "return business". As more employers sponsor employee training, more employees came to value continuing education.
- Workers face significant obstacles to their participation in training. A majority report that training is not offered at a convenient time or place; some believe it costs too much; some others say that no good programmes are available or that they lack the time and energy for training.

- Another important criterion for training, in the workers' eyes, is job specificity. Some cast a doubtful eye on adult education that is not directly related to career advancement; they want to be sure that training will pay off.
- Finally, workers express the need for support in meeting pressing family responsibilities. Many speak of the strain of needing two pay cheques merely to make ends meet, of having to care for children and of securing adequate health care, transportation and other basics.

Taken together, these findings suggest that employees will not respond to training opportunities without stronger motivation and help in overcoming obstacles. All of these factors combine to foster a core of cynicism in a substantial minority of the workers we surveyed. Frustrated by obstacles, skeptical of society's faith in education, ensnarled in red tape whenever they look at the educational process, many workers have lost faith in the system.

- **Awareness and Confusion**

This analysis adds up to a compelling argument for major changes in the way that government, business, education establishment and individuals operate. Workers need to dedicate themselves to the new reality that maintaining job security in the face of rapid economic change means constantly upgrading. Businesses need to understand that they must adapt to a new world order based on high performance operations that require significant investments in their workforce. The education sector must think about how to direct resources toward the learning needs of adult workers and to deliver more learning services at the worksite.

Most important, the analysis speaks to authority because this new information requires a different way of looking at government, a new paradigm, a fresh model to define state functions in a new economic era. State policy makers have to look at human capital as the next frontier in economic development. They have to transform narrow, targeted, programmatic approaches in education, job training, and human services into broader, "wholesale" initiatives. They have to infuse a new vision of what economic and human development might be into every nook and cranny of the state so as to reach and influence the greatest number of individuals and companies.

Above all, state officials have to reach out to the public and private sectors throughout their state with a new message: economic development at its best represents collaboration for the common good, a collaboration among public and private leaders to create high-skill workplaces and the highly-skilled workers that make such workplaces possible.

12.3.1 An Agenda for the Millennium

The need to change is clear.

The "system" must also change in response to the new realities described in this document. The key to our national progress, everyone has agreed, is education and training; but now these must be reformulated to meet the new demands of a different workplace. If we are to advance as a nation, high-skill, high-wage careers must be within the reach of all. All Americans, whatever their background, are entitled to the prospect of a future unfettered by the poverty or deprivation they may have experienced as children. Translating the appeal of that vision into the stuff or everyday life is this country's enduring economic challenge.

It has become abundantly clear that America is at risk if it simply accepts the underlying economic and occupational changes that are shaping the future. American companies are losing the chance to grow not just because of inadequate worker skills but because of their inadequate investment in improving skills. Many communities, many companies, and key state leaders still need to be convinced that education and skill improvement is the key to economic growth. But no major sector of society is investing adequately in helping people work smarter. Many in government and the private sector–by inclination, reflex, or force of habit–still hope to get by with business as usual.

There is a need for a new paradigm for economic development that actively promotes education and the polishing of workplace skills. It must realize that high-skilled jobs are the dividends paid to investments in people. It must reflect local needs and draw on local strengths. It must bring public and private sectors together into a common enterprise to build a better future. Above all, it must be designed to keep the jobs that we have today while creating better ones for tomorrow. Once developed, this paradigm, this new agenda for economic development, needs to be promoted on a vast scale.

A New Agenda

A new agenda has to be based on the economic realities in which we find ourselves today. The agenda must be grounded in ambitious principles designed to point youth, workers, and communities toward a different future. Any economic development plan not organized to transform that truth into operating reality is seriously flawed.

Objectives of the Agenda

(i) Economic Development Depends on Human Development

The first principle is the foundation of the new agenda. Adherence to this fundamental belief must become second nature to state and local officials interested in economic development. They must come to understand that economic development is not bricks and mortar, not fancy new technology or even trendy new management styles. Economic development starts and finishes with the people in the community.

To be improved, human investments must be seen as developing and maintaining an integrated system for learning, not as disaggregated pieces–public and private, elementary or secondary education, and higher education. States and communities must broaden their views of what needs attention and, in doing so, expand their notions of integrated workforce development systems.

It is important to know how well schools are doing. But if 70 or 80 per cent of state's future workforce is already on the job–and if 60 to 70 per cent of the state's high school graduates are migrating elsewhere in search of employment–then it is also important to know how well the adult education and training system is doing.

More than anything else, states need to begin seeing as a basic resource. The traditional view of pouring more funding into existing educational institutions does not serve well, the reality is that a state's future will depend on those who are adult workers today. In response, states, businesses and education institutions must begin to use the workplace as a learning place, to deliver learning where the people are.

(ii) Emphasize Lifelong Learning

One key to future success will be the development of public and private sector partnerships with a broader view of necessary human investments. Most states invest a majority of their education funding in traditional schools and institutions of higher education.

Schools and colleges are important. But other needs are also important to economic competitiveness. The targets of these other investments are the people outside of the **K-12** range, outside of the colleges and universities: those in early childhood, those who need to learn in the workplace and older workers who are in need of retraining. Responding to these needs can pay tremendous economic and human dividends.

States need to have a system of lifelong learning–a system of education that meets the needs of individuals and workers from cradle to grave, a seamless web from the perspective of the user that allows learners to move in and out easily as their education and training needs change.

States need to refocus educational systems to engage the majority of the workforce, present and future. Today, 50 per cent of our students who are not college-bound have a few credible options to develop the skills that technology-minded employers need. The system largely abandons dropouts; it offers little in the way of an integrated approach to the transition between school and work; it reaches adult workers in a limited, haphazard way. As a result, many companies do not use or trust public institutions to help their workers.

(iii) Recommendations

With these principles as a guide, outlined below are nine major recommendations toward which the state economic policy should be directed.

- **Development of an Integrated, Market-Driven System of Lifelong Learning**
  Authority need to undertake a variety of initiatives to make education and training more accountable, more able to provide a closely integrated lifetime of skill enhancement and more sensitive to the needs of employers and workers.

These would include such actions as:

- Strategic planning for education and training at both the state and the job-market levels;
- Better articulating the transitions from high school to work and from high school to higher education;
- Making use of the learning that takes place outside the school system;
- Development of an aggressive public strategy to bring the private sector into the market, demanding more of public institutions through consortia (association), financial incentives and public-private partnerships;
- Development of an aggressive public-education campaign, with public and private leaders explaining and demonstrating how education and skill enhancement lead to better jobs.

- **Development of a System that Makes Existing Institutions Accountable to Serve Workforce Needs**

> It is time to introduce the principle of accountability to the workforce preparation system. Among education and training institutions and public job training programmes, it is virtually impossible to measure what we get for what we spend. Public education budgets reward process but not performance; i.e., they pay for enrolments but not for learning or skills acquisition. There is no "consumer information" for employers or employees to compare the performance of institutions and programmes.

Further, the educational and training resources available are diverse, apparently unrelated and confusing. Many people and companies become frustrated by the complexity of the system and abandon public education or turn to private sources for help.

There is an urgent need for:

- State and local clearing houses or information systems capable of providing uniform, usable information about economic change, skill needs, job opportunities and training requirements at the labour-market level;
- Providing information to small business owners about literacy resources and other training programmes available to their employees through the local education and training system;
- Development of common measures of performance for institutions and the establishment of sanctions for non-performing institutions and programmes;
- Making performance information for education and training providers accessible to employers and workers.

➢ **Encourage Work-Based Adult Training**

Many companies have taken pioneering steps to invest in their people to achieve continuous improvement. Several action steps suggest themselves as ways to cash in on the tremendous opportunities for accelerating the amount and quality of work-based learning:

- Secure the commitment of top management, making training part of the company's plan;
- Find ways that allow the private sector, as the chief beneficiary of new training arrangements, to pay most of the cost of the training, using public investments strategically as "leveraging incentives" like tax credits and loan pools;
- Develop the ability of public sector institutions to train at the worksite during working hours;
- Keep education closely linked to work tasks;
- Design and encourage training for front-line workers, not just for management and professionals;
- Ensure that training is continuous, not just a one-time affair;
- Encourage companies to demand more of local educational institutions; and
- Offer incentives for the development of small-business consortia (association).

➢ **Improve the Transition from High School to High Skill Careers**

One conclusion from this research is that communities and states cannot effectively pursue new human resource investments without addressing the needs of the "forgotten half," those high school youngsters who are not planning on attending college. States and private companies must achieve a closer integration of school and work, between academic and occupational learning, and between classroom and work-based learning. One promising model is the creation of structured work-based learning opportunities for young people beginning with their last years of high school and extending through the first two post-secondary education years.

Such programmes should seek to build on the best practices of our European and Asian competitors who have well-developed career ladders. Yet they must also provide the flexibility and personal incentives that American students demand. A youth initiative provides an easily understood means to transform the learning system radically, motivating young people and linking them to employers and higher-education opportunities. It also ensures that technical training will enjoy the same prestige as college-track schooling and offer the same opportunities and rewards.

These school-and-work transitional programmes should seek to do the following:

- Arrange employment opportunities in key local industries with complementary in-school learning;
- Provide credit and credentials for work-based learning;
- Ensure that students in technical areas are prepared for advancement toward higher education if they so choose;
- Encourage achievement early in high school; and
- Enhance the status of vocational-technical schooling.

➢ **Close the Emerging Skills Gap**

One of the significant economic realities in any community is generally not well understood or acknowledged. It is the existing gap between the skills businesses need and the skills available in the workforce. It is an inherently moving target, with skill demands changing with the advent of new technologies, new markets, new products and new ways to organize within a firm.

This gap is never fully identified in any standard economic and employment data. The gap is usually only clear in hindsight (retrospective), meaning that policy-makers end up responding to skill needs as they were and not as they are today or will be in the near future.

Awareness of such hidden skills gap gives policy-makers a leg up in constructing a new education and training agenda. It calls for a detailed examination of private sector skill needs in light of what the public sector is providing. It calls for new relationships between the private sector and public education and training institutions and programmes. It calls for new ways of helping workers understand the role of an objective skill assessment in helping shape an education plan to achieve economic security.

By defining what is needed in the workplace and what is being provided by public and private trainers, analyses of the "skills gap" galvanize employer interest in issues of human resource development. It is a critical first step toward developing and maintaining the kind of skilled workforce a community needs if it is to meet competitive challenges home and abroad.

➢ **Involve Citizens in the Debate over Solutions**

At each stage of its development, the United States has placed its faith in the inherent good sense of the average citizen. The problems of maintaining economic growth and standards of living for this generation and the next are a central concern today of the average citizen. They are entitled to inform the thinking of local and state leaders on these matters. Even more significant, local and state leaders need to understand public thinking on these issues as well as how citizens can be involved in charting a course ahead.

Involving everyone in the process permits policy makers to emphasize that everyone has a stake in the outcome as well as a responsibility for making the desired outcome real. This presupposes building a constituency for change at the grass roots level, using the language and understanding of citizens to pave the way for the change process. People must understand what is happening in their economy and they must understand that they alone can create the necessary improvements.

Lasting and effective change requires strong "bottom-up" pressure, but it also needs visionary leadership to pull the process along. State leaders need a broad base of support if they are to help change American life, and the people at large need a challenging vision of what they can accomplish.

➢ **View Workforce Issues through the Lens of Substrate Regional Economies**

Workforce skills must be understood and met in the context of operating labour markets, not in the context of theory or state-wide data. Examinations of regional information in relation to each state lead to the conclusion that policy-making has to be sensitive to dramatic substrate and regional variations. Policy developed for **Colorado's Urban and Suburban Front Range** will have little relevance on the Western Slope; initiatives developed to respond to the needs of **St. Louis** almost certainly will not work in **South-Western Missouri.** Concepts applicable to **Rural Indiana** and its comparatively healthy small towns cannot be transplanted to **Rural Mississippi**, which has a far different set of needs.

Effective strategies must include:

- Development of a local capacity for economic development and workforce preparation;
- Creating new community and regional forums for strategic planning of human investment and workforce preparations decisions and for coordinating institutions and resources;
- Designing programmes that are managed locally and not by state-wide bureaucrats.

Job-training strategies in particular must be geared to the realities of the local labour market. Essential to this is the creation of flexible funding and innovative programmes to shift attention to the differing needs of states, regions and communities.

To be fully effective, this "localization" of resources and control must be accompanied by strategies that require participation by the users of the programmes, for example by individuals, local businesses and by communities.

In other words, in this new conception of government, public investments must be structured to leverage private investments. Rather than a system of entitlements or local aid, they must become a bridge to self-sufficiency and responsible initiative. The message should be that "we invest in you if you will invest in yourself".

> **Build the Demand for Higher Skills**

Many of the other recommendations that appear in this report focus on ways to build a structure to develop higher skills within the workforce. States must give a high priority to building the demand for higher skills within their state economy, as well. Putting it in another way, the states must give the highest priority of their economic development to modernizing the existing industrial base.

Many companies around the country have shown that they can restructure themselves into high-performance organizations capable of competing on a global scale. These businesses combine the latest in production technologies with innovations to make the highest quality goods and services. The strategies to accomplish this, all emphasize, quality and the continuous improvement of worker productivity.

The strongest demand for new and higher-order skills comes from these companies. But comparatively few businesses have been attempted to adopt a high-performance profile. State economic development in the nineties must help build demand for higher skills by encouraging more existing industries to adopt a high-performance posture. States can:

- Assure that they have technology transfer programmes in place that help small and medium-sized companies gain access to available production technologies;
- Encourage groups of companies–business or trade associations, or several companies located within a single community–to work together to identify and solve common workforce needs;
- Make sure that all state business development programmes incorporate a component for human resource development;
- Encourage larger companies to promote technology assessment, management assistance, and training along supplier networks (larger companies have the internal resources that their smaller suppliers lack).

> **Attack Workforce Challenges on a Scale Large Enough to Make a Difference**

Most state interventions, whether job-training programmes or manufacturing modernization, affect only the smallest tip of an enormous iceberg.

The public sector can undertake a variety of strategies to leverage significant or dominant investment and ownership in training from the private sector. Many kinds of initiatives can reach and involve large numbers of people if they are created, funded, and owned largely by the private sector. Small-business training consortia, apprenticeship programmes, dropout prevention, modernization efforts and many other effective programmes can be convened by state leaders and encouraged by the strategic use of incentives.

A critical component of this effort is the need to confront the misplaced notion that workers who need retraining are somehow "failures". The private sector can be the public sector's most effective representative on this issue arguing that training–by adding to skills–adds to workers' income, their value in the company and their importance to the community.

A New Policy Process for the New Agenda

Assuring progress in responding to economic transition on the scale that is clearly needed suggests undertaking of a new public policy process. Three realities are clear.

- First, achieving the new agenda will be accomplished, in part, within the context of the existing political climate. Traditional political debate over goals and over the allocation and investment of scarce resources will continue. The situation in the states today is one where extraordinary pressure exists to meet short-term needs with fewer resources. It is not a climate that encourages significant innovation.
- Second, solutions will require an unprecedented degree of collaboration between sectors, between business and labour, government and private sector and between state and local agencies. For example, business, labour and education should all work together to prepare an integrated curriculum for new youth transition programmes. Without formal collaboration, the goals of learning at and for the workplace will not be met.
- Finally, implementation will involve maintaining alliances and relationships among groups with fundamentally different political imperatives—management and labour, corporations and small business owners, the profit and non-profit worlds, traditional educators and adult trainers—over the long-term. There are no quick fixes. States can no longer hope that limited action at the margin of public policy will have the desired effect. New structures must be created both to institutionalize the ability to respond quickly to constantly changing conditions and to remind the diverse actors of their common stake in success.

These are the new rules that states should follow in defining the process to accommodate change:

- **Take long-term view.** Economic transition creates challenges without regard to the terms of office of state leaders or budget and legislative cycles. States should seek to institutionalize a process that is able to withstand changes in political administration and avoid getting caught up in biennial (yearly) debates over funding based on changes in the state's fiscal position.
- **Keep an open door-admit into the policy debate of a wide variety of sectors and interest groups.** Advisory boards should comprise leaders from business, labour, education, community organizations and state government. They should also include key representatives across several state agencies as well as the state legislature and attempt to balance state-level with local-level leadership, including private citizens.
- **Nurture new relationships.** These advisory groups offer a way for leadership from different sectors to work together on common issues. They should use their power to build new and permanent collaborations that are part of more effective multi-sector approach to public policy.
- **Give the process high visibility within the state.** In almost every case, appointments to multi-sector advisory groups should be made by the Governor of the State, who should

demand a clear set of recommendations for action. Regional meetings, citizen meetings, media publicity and other techniques should be used to build awareness of and support for new policy initiatives.

- **Inform the "process" with appropriate strategic information and choices.** This report points out the need for states to collect and synthesize new kinds of information that is not otherwise easily available. This includes diagnostic information about the nature of economic change and the skills implications. It means knowing a lot more about the outcomes of investments that the state already makes in a variety of programmes and institutions. And it means sharing success stories from elsewhere in the state and the nation that may have applicability in the local setting.
- **Make participants feel responsible for implementation.** The action plan that results from the policy debate will challenge every sector to behave in new and different ways. Participants in the policy process must include leaders from a range of groups willing to commit themselves to promoting the new action plan among their constituents.

A Better Tomorrow

These are trying times for the American worker. Foreign competition and wrenching economic change in the 1980s promise to accelerate in the 1990s as Europe prepares to unite, newly industrialized countries continue to advance and energy costs threaten to rise.

- Now Indian resilience is being tested anew. But Americans have always known how to stop, how to change directions and how to channel their energies into productive new directions. We must do so again.
- It is time state leaders stopped and framed the need for change in their communities in compelling new ways. The research for this project indicates that the public will respond-that the public is, indeed, eager for an authentic restatement of traditional American values in the context of the economic demands of today.
- It is also time we changed direction. We must stop thinking of employment and training issues as isolated components of the nation's economic dilemma and come to understand them as part of a comprehensive whole. The traditional approach to dealing with social problems in the United States has tended to be fragmented and piecemeal. It has left behind a legacy of disjointed, but well-intentioned programmes. If we are to clearly gauge priorities and understand interconnections among different pieces of the economic development puzzle, communities must integrate individual views, business impressions and public data into a unified whole.
- Such integration, of course, is precisely the responsibility of leadership in a free society. It is the most difficult–and therefore the most important–task facing those who would channel public energies into productive new directions.

Managing the Indian workforce is unique in many ways. Some of the theories put forward and many of the practices adopted in developed nations do not work very well in India. For example, Table 12.3 shows India vs Japan in social and cultural factors.

A certain transformation of the workforce has to be carried out before these theories and practices can be applied effectively. The reasons for this are embedded in the country's culture and traditions – the way its people look at life and live. These cultural and social characteristics and traditions also affect the way people behave. Therefore, in order to be able to manage the Indian workforce effectively for agile manufacturing, one needs to understand the Indian social and cultural environment and its influences, since these give rise to various problems which managers of Indian manufacturing companies need to face. Rampant indiscipline, feudalism, bureaucracy, and negative, or 'soft' work culture are prevalent in many of the country's industrial establishments. These have resulted in the lack of a sense of belonging, poor productivity, inadequate quality of products and processes, an erosion of profit margins and an insufficient rate of growth of the country's industrial establishments in particular and of the Indian economy, in general.

Another indicator of the work attitude and the prevalent work ethos in India is the number of workdays lost.

TABLE 12.3: Social and cultural Factors–India vs. Japan

Social/Cultural Factors	*Japan*	*India*
Principles of government/ Administration	Emphasis on government by the virtuous and abrogation of coercion; mutual trust between employers and employees and acceptance of the basis goodness of human nature. The result is minimum control from above, a high level of delegation, and a highly motivated workforce.	Prevalence of impersonal bureaucratic social relations, and a system of administration based on mistrust of fellow beings. The result is a highly centralized administration, over-emphasis on hierarchical status in decision-making, bureaucratic delays, a dissatisfied workforce, and an accentuation of apathy in individuals.
Attitude towards work and goals	High result-oriented and directed towards perfection and growth through dedicated effort.	General apathy and dissociation of work from its results. Tasks are performed without any interest, dedication or pride.
Discipline and order	Highly disciplined; respect for superiors and for authority.	General lack of discipline; mistrust of authority, and poor superior-subordinate relationships.
Group harmony	Very high; based on informal affiliative pattern of behaviour.	Assumption of the inequality of human beings; nagging suspicion of fellow beings and highly self-centred behaviour resulting in a lack of cooperation and teamwork.
Emphasis on education	Very high.	Generally indifferent and highly ambivalent. Education seen by the common man only as a means of earning a decent livelihood.

12.3.2 Some Problems of Indian Workforce

The primary reason for such a poor performance is the country's apathy towards work and work-related activities. One of the indicators of this is the average number of hours worked per week in the India's manufacturing and business establishments. There is also a significant difference between the hours worked by workers in shops and the hours put in by office workers. Whereas workers in shops are required to work 40 hours per week (5 days), the average number of hours worked by those employed in offices is only 33. This shows that office workers are more privileged.

12.3.3 Why Indiscipline in Indian Workforce?

- Lack of interest in allocated tasks and excessive socialization during working hours thereby disturbing those who want to work
- Excessive specialization of the individual during work hour thereby disturbing those who want to work
- Dirty surroundings and poor housekeeping and unsafe work practices
- Lack of personal discipline and ignoring orders, aggressive behaviour, unauthorized absence, wilful tampering of instruments and theft of company property
- Overstaying, lack of organizational discipline, high absenteeism, claiming over-time payment, often shifting workplace causing loss of production
- Militant and negative approach, tendency towards unionism
- Working short hours, disparity in work between individual and group
- Overemphasizing, claiming hierarchical status and accepting too much responsibility levels although he cannot discharge
- Lack of participation and absence from participative forum
- Casteism, distressing, categorizing and boosting superiority as belonging to superior groups
- Legalistic and assertive approach in expressing grievances and asking for welfare measures
- Disinterest casual approach and reluctance to obtain education training
- Not getting on well with high-ups and superiors, getting demoralized and alienating managers
- Fragile self-esteem, lacking of confidence, age-old arrogance and boast fully claiming high intellectual levels
- Lack of trust in seniors and superiors, nagging suspicion about other people
- Lack of gratitude for help received and intensive reluctance to meet out disciplinary measure and punishment when awarded

12.3.4 Job Enrichment

Changes in job content will only be acceptable to employees where such changes are treated as opportunities rather than as unreasonable demands imposed on them by management. Therefore, for work innovation to be effective, it must be brought about only through participation.

The following are some of the more important measures to be taken for developing a positive work culture.

- Be fair but firm in dealing with subordinates
- Take active interest in the personal and family needs/problems of subordinates
- Discourage overemphasis on hierarchical status and 'pulling of rank' in day-to-day dealings
- Promote information sharing and communication through periodical departmental meetings, production committees and small group activities/quality circles
- Promote the setting of goals/targets and decision-making through consultation and participative forums and ensure consistency in decision-making.

India is a very large country with a wide heterogeneity of languages, ethnicity, cultures, attitudes, ideologies, preferences and economic status. Hence, these diversities must be recognized and honoured; close-knit human groups and communities take pride in their own distinct entities and dislike being taken for granted and told from above as to what is good for them. Thus, the development of a national work ethos requires coordinated efforts on many fronts and by many agencies, such as the government, schools and colleges, technical institutes and professional. This is a complex and difficult task, but it must be pursued relentlessly.

12.3.5 Stress on Training and Re-Training

In business, growth is necessary for survival. This is particularly true today with the liberalization of imports and increasing competition. Moreover, growth and higher productivity create employment and higher employment creates larger markets. All this planning needs a high level of technical competence and professionalism.

A professional requires a high level of training and years of experience to become competent in his chosen occupation – be it engineering, accountancy, marketing or industrial relations as shown in Fig. 12.10. It is obvious that good professionals will turn out to be good managers and have a much better perception of the constraints and variables on the end result they are to achieve. Technical and professional competence is essential for an objective evaluation of alternatives and for proper decision-making.

Everyday, in India one comes across instances of exploitation of the weak by the strong; this is brought about by the feeling of inequality between men. Inequality and exploitation have tortured Indian society for far too long, dividing it and making it weaker. At the same time, bureaucracy and feudalism (social systems) have caused a lot of harm to the country's organizations. This has resulted in a loss of product quality and a demoralized and alienated middle management. These conditions must be set right at all cost.

Disobedience of the orders of superiors, aggressive behaviour, unauthorized absence, wilful tampering of instruments and controls bordering on sabotage and theft of company property as explained earlier are examples of lack of personal or individual discipline in the workforce. These are indeed destructive and stop progress.

Indiscipline has a very deleterious effect on employee morale, on production operations, productivity, product and process quality, on plant and equipment availability and on other indicators of organizational performance. It also affects cohesiveness in the workforce and its sense of belonging to the organization. Lack of discipline is disease – a condition which is very difficult to correct and can only get worse.

Discipline should be ingrained and imbibed even during early years by the workforce in India. Inculcation of discipline in the workforce is and will always be, an important responsibility of organizations.

The most important measures for positive work culture are:

(i) High level of group harmony in all functions, divisions, departments and sections based on a strong sense of belonging to the group and a feeling of caring for each other.
(ii) Parity of authority and responsibility at all levels of management

(iii) Management by the virtuous people with honesty, sincerity and impartiality in the management of the organization (including day-to-day operations)

(iv) Superior-subordinate relationship based on respect and a sense of reciprocity and mutual dependence. On the part of the subordinate, the relationship should be devoid of servility and conscious efforts at self-assertion; on the part of the superior or senior, it should be devoid of autocratic behaviour and feudal (social) undertones

(v) Accountability for performance-quality, efficiency and effectiveness of the work done–in all functions and at all levels

(vi) Rules and regulations based on trust and strict adherence to set systems and procedures, with their rules and regulations

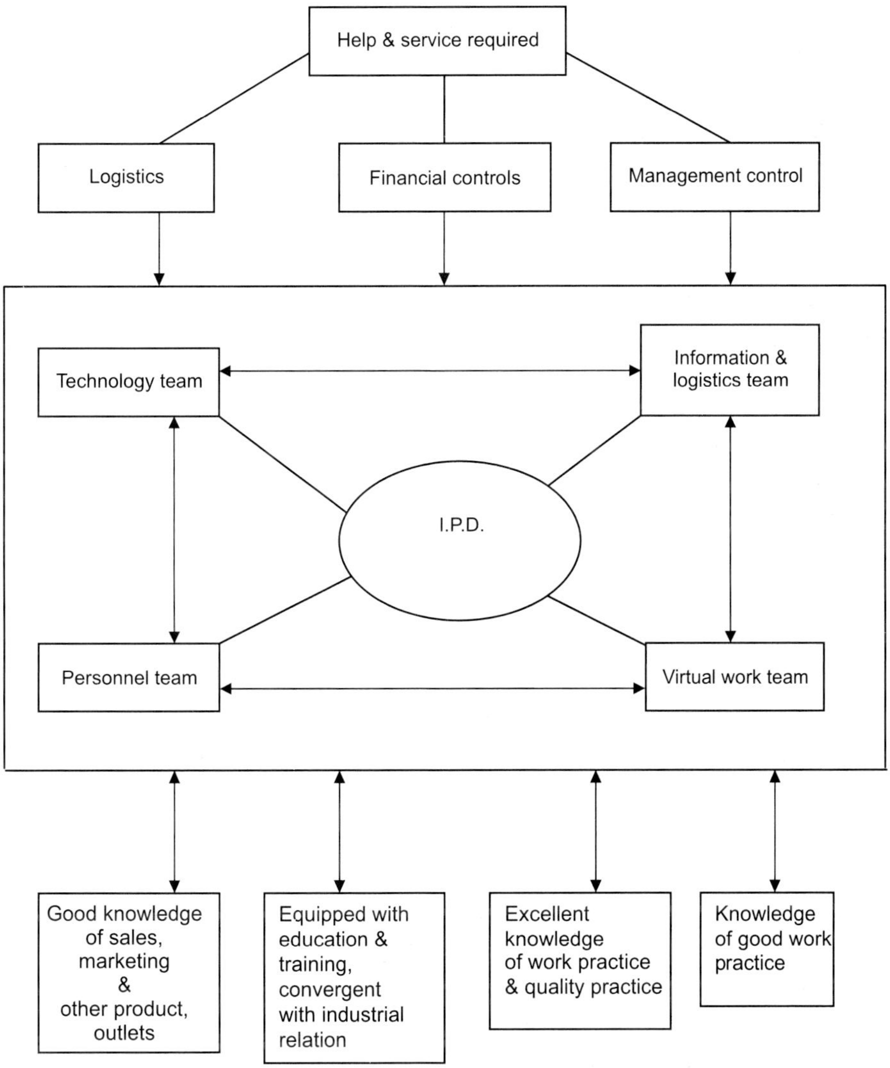

FIGURE 12.10: Establishing integrated product development (I.P.D.).

(vii) Abolition or minimum use of force to get work done and emphasis on self-discipline at all levels
(viii) Strong emphasis on education and availability of avenues for self-development
(ix) Prompt identification of unscrupulous elements and weeding out of hopeless and corrupt employees
(x) Strict enforcement of discipline in all functions and at all levels of the organization.

To Sum up

In addition to highlighting the affliction of high absenteeism and high overtime, two other indicators of organizational indiscipline are loss of production during shift change and lack of attention to safety. Undesirable and dysfunctional stress on hierarchy and hierarchical status, in turn, is a derivative of bureaucracy. Creation of structural groups and domination by some groups or relative importance of some groups is also a derivative of feudalism (socialism).

The discussion so far has indicated clearly the various problems concerning the Indian workforce. Some of the problems to repeat are: the dysfunctional elements and the stumbling blocks. This, as one realizes, calls for transformation – a radical change in the way Indians view work, rules, regulations and procedures at the workplace, their fellow beings and co-workers, and the basic values of honesty, perseverance and teamwork. But transformation requires action, since knowledge by itself, does not beget action. For this, a strong dose of conviction is required and conviction comes from confidence and trust; confidence in oneself and one's abilities and trust of others–including colleagues, fellow workers and management. Confidence comes from motivation and self-discipline in individual employees and the workforce is, in turn, a function of empowerment. Therefore, the pivotal factors in bringing about change are motivation and empowerment.

> **Note:** Such participation is important for the workforce in question to intimately learn practical working methodologies in real situation and enhance progressively their skill and confidence.

A diagnosis of the performance of employees in our work environment shows that poor performance is caused either by execution deficiency or by motivation deficiency. In other words, in some cases, employees want to perform well, but cannot because they do not have the necessary skill, knowledge, trained ability or the wherewithal.

Motivation and training are an essential part of leadership and leaders & managers must be capable of motivating their subordinates because they have the unenviable 'task of creating a true whole that turns out more than the sum of resources put into it' **(Drucker, 1975)**. Motivation is the process of stimulating active interest that starts with a need or felt deficiency, either physiological or psychological. This need activates the behaviour or drive, directed at a goal or an incentive. The need gives rise to the motive and the primary job of the leaders and managers is to awaken the need in their subordinates to achieve and excel. This leads to:

(i) Getting specialists, engineers, managers, etc., for self-help and to act as implementation agents – Fig. 12.11
(ii) Choosing partners to help themselves.

The basic inputs for a conducive work environment are:

- Employment security and safety at the workplace
- Clean, tidy and well-ventilated workplace
- Unbiased and fair evaluation and periodic feedback of performance
- Recognition of a job well done and of good performance
- Encouragement of innovations and suggestions for improvement
- Award for excellence – for performance of a very high level and/or for good implementable ideas/suggestions for improvement in productivity, quality, equipment availability and effectiveness

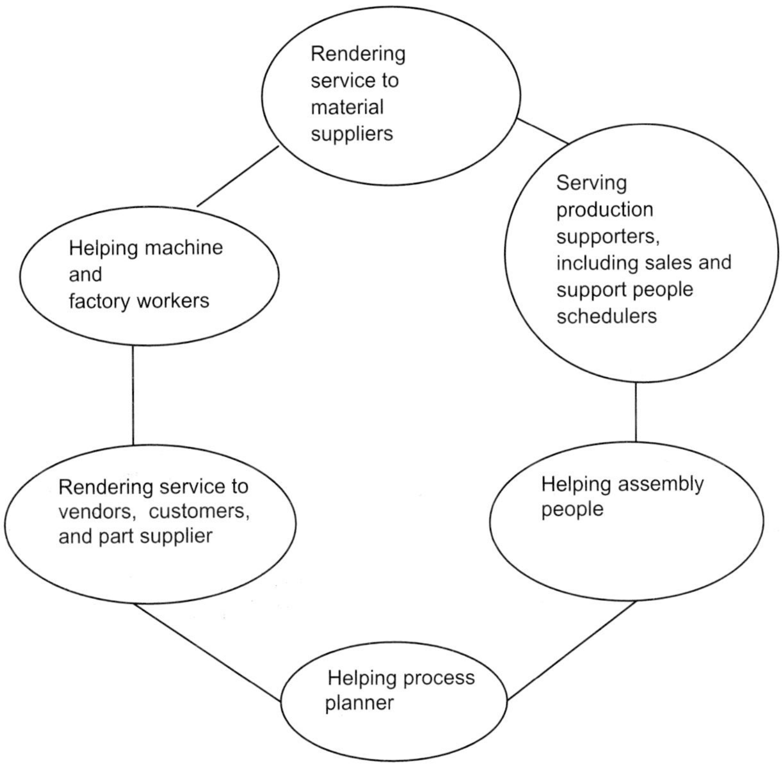

FIGURE 12.11: Getting specialist engineers, designers, manager, etc. for self-help and to act as implementation agents.

12.3.6 The Changing Face of Manufacturing in India

Till few years ago, the customers' main concern was the price of the product manufactured by the organization and thus the basic objective was minimization of the manufacturing cost through the maximization of efficiency. This led to economies of scale and focused production systems. However, together with a reasonable or competitive price, the customer is now calling for greater product variety, and improved product quality. This has given rise to the need for flexibility, or what is now known as economy of scope, which entails quicker response to customer demands, products and process optimization, process control and continuous improvement. Economy of scope emphasizes the ability of an organization to cope gain with variety. The greater the scope, the greater the economy it can gain from reusing and reapplying its intellectual and physical resources to the different types of products it produces (Busby, 1992).

Four decades have gone by and the country has still not been able to either banish poverty or build a nation which is truly prosperous and strong. The population has grown manifold during this period and this has made the tasks of reducing poverty and nation building even more difficult. Rapid industrialization by itself will not solve the problem since India's existing industries are plagued by poor productivity, inadequate product quality, late deliveries and high manufacturing costs. These industries have to become more competitive and grow instead of seeking protection and subsidies from the government. For this, they will have to imbibe new technologies and new manufacturing processes, become more flexible and improve their productivity and reduce manufacturing costs, ensure high quality products and services and improve deliveries. Most of the existing companies will also have to redesign their business processes in order to achieve competitive capability.

Introduction of New Technology

These new technologies include not only advanced manufacturing technologies (hardware), but also advanced information technology tools for design, scheduling and control (software). The applications of the new technologies calls for a new breed of engineers and technologists, who are equally adept in functional knowledge areas as in the use of computers and computer programming, methods for chain supply (of products), etc.

The need to modernize through the implementation of agile manufacturing was further accentuated by liberalization and the opening up of the economy in 1991. This was a major change for the Indian manufacturing industry since it opened up a protected economy and many manufacturing organizations found themselves threatened by foreign competition. However, the Indian organizations have, by and large, responded vigorously and positively to the changes in economic environment (Nilakant and Ramanarayan, 1988). The larger manufacturing organizations have taken up the job of modernizing their manufacturing organization, manufacturing facilities, processes and procedures. This, however, brings up a very important issue. Although it is generally recognized that the modernization of industries is essential to improve our productivity and reduce costs, the fact that the human factor is the key issue in modernization is not so well appreciated. The effective use of human resources is a critical factor and the implementation of new technologies calls for an in-depth examination of the existing management structure and the ways and means of getting the work done, the reasons for the lack of innovation and steps needed to create a conducive environment, use of staff specialists, role of middle management, recruitment, training, performance appraisal and career development of middle managers and white-collar specialists, and the role of employee unions and associations.

Need for Teamwork and Use of Empowered Teams

The individual is considered the basic organizational unit and practice of management stresses the perfection of the technical skills of the individual and assumes that this will lead to an effective and productive organization. **(Scully, 1996)**. Scully suggests that instead of focusing on the individual and his technical skills, managements must stress the improvement of the effectiveness of groups of individuals which make up the whole organization. This is because organizations consist of employees and working together can achieve great things because of the synergism created by the bringing together of diverse skills (Hradesky, 1995)

Moreover, there are many people who cannot manage everything on their own and for such people, empowerment is not enough, since they need the support provided by the other team members to be truly productive. Therefore, the most effective organizational unit is a team of individuals which has been given clearly defined goals and also adequate resources to accomplish these. Moreover, the team should be empowered in such a way that team members feel a sense of ownership of the work as well as a sense of belonging to the organization.

The success of a team depends on the team members. The team members must learn to participate, communicate, respect one another, exchange opinions, consider all the options while voicing an opinion or coming to a decision, etc.

The factory of the future is within sight, and it will be here in India sooner than we expect. The country should prepare itself for it – both technologically and socially and psychologically (attitude-wise)–as otherwise it will be caught unawares resulting in a lot of pain and suffering. Its schools and colleges, its social institutions and the government must prepare its young citizens – the workforce of tomorrow – to face this challenge.

12.3.7 The National Challenge and the Need for Progress

Following World War II, the people of many nations began to take serious steps and took great pride in note of their nation's place in the world economic pecking order. They found the tables had started to turn, overseas competition had started draining manufacturing jobs to other shores. Services join production as a cornerstone of any economy. The electronic revolution simultaneously simplified and complicated even the most mundane tasks. For the first time, each nation had to race to keep up with economic evolution.

It is outlined here how major changes underway in the national economy play themselves out and points out how governments, employers and employees are not yet responding adequately to these changes and how many key leaders are playing a new game by using old rules.

The Imperative for Change

Change, as the saying goes, is the only constant. Every country today faces changes that are unprecedented in scope and implication. People expect technology to accelerate, but as technology is changing, it is sweeping away old ways of doing business and in many cases, peoples' jobs. Innovations in communication, manufacturing, and transportation are making businesses more competitive, but these innovations give equal advantages also to competitors. While workers are expected to set the standard for the world, they are found to lack the skills required in the modern workplace.

As a nation, many of us have not yet mustered our common strengths to confront these changes. Awareness of the situation is certainly growing, with many state and local officials eager to take constructive action. The population as a whole is beginning to get a clear fix on tomorrow's horizon but is uncertain how to interpret local change within its worldwide context. Business leaders talk about the need for new skills but invest very little in their workforce. Despite focus on the declining quality of education, the response lags far behind the rhetoric (talking). Although states seem willing to change, the work of this project suggests that communities and states are only beginning to mount an adequate response to the economic challenges before them.

Findings in General

The findings of this project reflect a picture of pervasive change in local, regional and state economies. National trends, those underlying economic and demographic shifts that touch every state to some degree, are actually affecting the states in very different ways: for example, somewhere urban service sector growth means expansion of high value-added producer services; in some other place, growth concentrates in retail and personal services; or manufacturing growth spawns high-technology industries; or somewhere else the challenge is one of modernizing basic industry.

Against this background, communities across the country are experiencing major shifts in employment by the industrial sector and many are experiencing shrinking demand for the kinds of jobs that created and expanded the middle class.

As a result, the nation's workforce continues to split into two disparate parts. A small minority, empowered by education, is highly skilled and highly paid. Others, isolated from supplementary training by employers and schools alike, are relegated to low pay for work requiring minimal skills. And because these issues of livelihood and productivity are central to life, their implications are interrelated and profound.

Sectoral Employment Shifts

The kinds of jobs available to the workforce have changed dramatically. The business consequences of the growing importance of the service sector are predictable. The human consequences are deep. Residents report seeing the effects of this trend around them, profoundly changing the lives of their families. People often express from their specific experience that "things are changing" and "jobs that are here today may not be here five years from now. And the company that has that job now has to have employees that can handle the new jobs".

The implications of the shift to services differ markedly, depending upon which part of the service sector is expected to grow in any particular place. In some parts, for instance, a large percentage of new jobs are expected to be in the service sector and half of them will be in producer services (these are jobs that provide services to other businesses and include such things as legal, accounting, consulting and marketing services). By contrast, in some other parts, even a greater percentage of new jobs will be in the service sector, but more than half of these will be in lower-paying retail trade positions.

Despite these distinctions, workers tend to identify service-sector jobs exclusively with lower-paying jobs like those available, say, as in fast-food restaurants. Discussing the disappearance of familiar mid-level jobs, people reject since service sector jobs do not match with available job skills.

Higher Skills and Greater Educational Requirements

The kinds of skills that people need to have–even if they hold the same job as before–have changed in the direction of higher skills and more of them. The competitive workplace of today–regardless of the product or service–is a high skill environment designed around technology and people who are technically competent. Assembly-line workers must now understand their work as part of a much larger whole. Many workers must be comfortable with computer-numerically-controlled equipment. Front-line supervisors must confront tasks like budgeting and fiscal planning; clerical and other support personnel must handle complex word-processing and spreadsheet functions. Relatively few of these workers are likely to have graduated from college, while not long ago most of these skills would have been demanded only of the college-educated.

But the problem runs deeper. It implicates the basic schooling that most people receive. There are widespread deficiencies in the basic skills of the recent high-school graduates working as labourers and clerical support. Many workers have difficulty in understanding the work environment and its expectations, even such basics as attendance and punctuality.

The "Haves" and the "Have Nots"

As mid-level jobs disappear, our society is dividing between high earners "empowered" in the workforce because of their high-level skills and those in survival-wage jobs, consigned to unskilled employment.

The middle class and the poor are threatened by income bifurcation. Unable to access school-based training and blocked by employers' inactivity in providing on-the-job training, these essential workers are unable to move into the upper half and unwilling to go quietly into the lower.

This income polarization is the one economic development trends that is obvious to respondents in all places. The growing gap between those making a good living and those struggling to survive and the difference in status between "professionals" and "low-skilled" workers, etc. is readily noticeable by the increasingly poor and rich people in the middle class sector.

Social Implications

The social implications of these three major developments are sobering. But these findings are not, in themselves, new. Other groups worried about national progress, state development and community growth have pointed articulately to the same set of issues for years together.

It is the lack of commitment to act on what knowledgeable observers agree needs to be done; that is one of the most striking and challenging features of this study. Researchers themselves ask: Why is it that so very little genuine change is apparent? The answer is complex. It is made up of several different strands.

12.3.8 The Causes of Our Failure to Act

Generally speaking, it appears that even if one feels the need to change, people are reluctant to act. Workers do not understand the changing economic scene and researchers note the shadow of the anxieties that always hinder human progress–fear of the future, fear of the unknown and fear of meeting failure.

But against these anxieties are people eager to keep pace with a changing world and to obtain information about how to do so. Politicians and researchers find political leadership wanting to experiment with solutions but frustrated by the difficulties of moving bureaucracies and communicating a coherent vision of a stronger future. Across the board, the single greatest obstacle is the need for understanding: understanding what kinds of solutions will work, understanding how to construct viable solutions, and understanding how to use the tools–the pragmatic innovations–to effect positive change.

Such understanding requires business leaders to act on their perceptions of the state of training, competition and the world economy to back their words with resources as they have never done before. It presupposes credible public leadership, able to communicate facts to its constituents, ready to build strategies in partnership with them and willing to share the burden of reshaping the workforce to the demands of a new economic age. For both political and business leaders, it requires beyond all else the ability to see the world from the workers' viewpoint, to recognize that most Americans lack clear and useful information.

Above all, scholars tend to conclude that there is an unbridged chasm between what seems to be obvious about economic change and what the public believes, between what needs to be done and what is being done. In regional meetings, people are alert to the sea changes of the world economy and eager to reassert a national position of economic leadership but uncertain about what these pressures really mean for their families, their employers and their communities. Part of this inaction, undoubtedly, reflects a general sense of powerlessness, which is most apparent in the focus groups and worker surveys. But much of the failure to act must be laid elsewhere, to the fact that such challenges can only be met through political action and political action is never unopposed, always diluted by compromise and effected only through cumbersome institutions. Add to these factors, a definite distrust of experts and leaders and the result is a nation poised for indecision.

A Sense of Powerlessness

Project findings confirm that people feel they have lost control over their economy. Governments find it difficult to respond to the situation and this is perceived by their constituents. Many existing governmental agencies are established in response to the rise of manufacturing; they are to answer questions that often are no longer asked. Created to foster "employment security," their very function is called into question when job insecurity is a defining characteristic of the economy.

Workers of today seem unable to sort out the barrage (large number) of training opportunities coming to them from all directions. With some exceptions, their appreciation of education is genuine and their desire for improvement is real, driven by the concern for economic change. But many of today's workers have let this concern incapacitate them. Participants from all four states agreed that they, personally, needed training but claimed that it was unavailable–despite an array of advertisements for literacy programmes, cooperative education programmes and community-college offerings.

This confusion is not surprising. On the one hand, employers do not offer sufficient training to help their own workers fill the more demanding jobs that open in their workplaces. On the other hand, many displaced workers see their own jobs are becoming obsolete and they are not aware of the new jobs that are replacing them. The nature of these new jobs–as well as the qualifications needed for them–are as mysterious to most workers as the hodgepodge (hotchpotch/confused mixture) of training opportunities available from an ill-coordinated mix of public and private institutions. There are many cases in point.

Confusion Over the Value of Training

Many of the workers with whom authorities have talked believe that they need more education and training both to keep the jobs they have and to get the jobs that they desire. By and large, they are not satisfied with the kinds of training they are offered by their employers. The workers surveyed want formal training that leads to a credential and to career and personal growth. Many would prefer this formal training to the informal, on-the-job training they most often receive. Many are sceptical about the words "job training" so commonly used by employers and representatives of the public sector. They view this as demeaning (lowering of dignity) and somehow related to welfare, a sign not of the economy's need but of the worker's failure.

Some people also face a long-standing bias against education itself. A considerable number of respondents deprecated the need for education in general. "I know people who can't read or write," said one man. "They make more money than someone coming out of college." "Who needs education?" asked another.

Even workers who value education do not know what kinds of jobs to look for or what kind of training to seek. Aware only that their training and experience are no longer as marketable as they once were, many workers simply do not have the perspective they need to direct themselves toward alternative careers that better suit today's economic climate.

Participants acknowledge the fear of losing their jobs, but no one can outline a plan of action to deal with that fear. They worry about the changing nature of work, but no one can describe career-development strategies that they would pursue. They speak jobs leaving their areas, but none can talk of relocating themselves in search of work. While participants are able to speak about economic problems in terms of their impact on their personal lives, they cannot outline a personal solution.

The Difference This Time

Devising new answers to new problems has always kept the **United States** at the forefront of technological development. The difference this time is that today our economic difficulties are not self-contained. In the parlance of the financiers of **Wall Street**, **the American** economy is in international play. **American markets** and jobs in sustainable economies will depend on investing strategically and learning and working smart.

12.3.9 How to Train Workforce Interested in their Jobs (Ref: Kennneth A. Tucker)

Steps for each of those:

- Workforce who love their jobs
- Workforce who are apathetic (indifferent) to their jobs

- Workforce who hate their jobs

(i) Workforce who love their jobs are more emotionally engaged than their less passionate colleagues. They are energized by their work; they keep learning and developing new ways to be more effective. If the job is so great that sometimes you have to pinch yourself to see if you're dreaming? Congratulations! Here are 3 steps to love the job even more:

- **Create ways to do more of what you're already good at and love to do**

If your situation allows, tell your manager what you particularly enjoy. Explain to him or her where your talents lie. Work with your boss to find opportunities to make more of your strengths. Volunteer to do more.

- **Develop a reputation as an "expert"**

You have special talents, skills and knowledge that others probably lack. Perhaps you are more strategic than your colleagues or you are a particularly talented communicator. May be you have knowledge and skills that your co-workers would find indispensable. Why not become a "go to" person who can offer others advice or counsel, drawing on your areas of strength?

- **Identify and partner with others who have similar talents**

Find your talent "soul mate". This may be someone whose talent is more developed or less developed than yours is in the role. Seek out someone who can offer you the inspiration and motivation that deepen your feelings towards your work.

(ii) Workforce who are indifferent towards their jobs are only minimally productive. They stay "under the radar," doing just enough to get by, but not contributing energy or enthusiasm to their workplaces. If you're bored or indifferent, do something about it–now, before it's too late. Think back to a specific time when you were excited about your current job. If it's been while since you felt that way, here are 3 steps how you can get the excitement back.

- **Renounce the present state.** You did not take this job to be apathetic. So if you find yourself watching the clock all day–or climbing the walls – admit to yourself that you're bored stiff. Declare to yourself that you want to make a change. Recognizing a problem is the essential first step towards solving it.
- **Remember your first day on the job.** What talents were you looking forward to using when you first took this job? Are you using them? How often? Are your talents being employed more than or less than, your weaknesses are?

Chances are, your talents are being used minimally, if at all. So think about how you could use your talents intentionally in your present job. What would have to change? The company's processes? Or would you, your manager or your fellow employees need to change? How could you transform your present role into dream job by applying your talents and strengths? Ponder over these questions and start taking action based on the answers.

- **Reclaim the joy.** Someone once asked: "Do I sing because I am happy or am I happy because I sing?" We don't know if there is a definitive answer to such a question, but there is one clear outcome – happiness. All of us want to feel happy and to be fulfilled in our jobs. So identify that special something that brings you joy.

For **Jason, a custodian**, the mere mention of clean window sills helped him to rediscover joy for his work. Even though **Jason** worked for a company that prides itself on hiring only the most talented people for every role, he once slipped into apathy. Cleaning offices, mopping floors, polishing fixtures, making restrooms sparkle – tasks that initially brought him a sense of achievement – no longer gave him satisfaction.

Then something happened that rekindled his joy. One day, a visitor to the company approached Jason with a broad smile on his face. The man, a job candidate at **Jason's company**, thanked **Jason** for positively influencing his decision to join the business.

Apparently, the man had doubts about the company and he set out to discover its flaws. But he just couldn't find any. People did their jobs incredibly well and seemed happy at work. Finally, in frustration, the man thought of an extreme test: "I'll bet you," he said to himself, "their window sills are every bit as dusty as other buildings' window sills," He promptly ran his finger over a window sill as he walked through the hallway. To his surprise, it was spotless. "And that," he said to **Jason**, "was when I decided to join this company."

Jason admits that there are still times when cleaning is anything but exciting, but that memory has kept him from feeling apathetic about his job ever since.

(iii) Workforce who hate their jobs – employees who are "actively disengaged." They're absolutely miserable; they cause trouble at the workplace as they spread their negativity to others. If you despise your work, get out! But wait: Before you go running off to land another job that you will most likely hate yet again, here are some questions to ask yourself:

- You chose it, after all. Think about why you chose it: Were you chasing more money? Status? Were you looking for more training and experience? Did you want the opportunity to become more well-rounded? Any of these reasons could lead you to hate your job.

More money is appealing only until your lifestyle catches up with your new earnings – then you crave even more. Status has its limits – what good is it to brag about your lofty job little or big-name company at a cocktail party if you're actually miserable at work? Training and experience are beneficial only if they help you develop your natural abilities. Well-roundedness is more likely to produce mediocrity than excellence; if you're well-rounded, you will become OK at many things, but great at none.

- **What feeling do I lack in this job?** Chances are, you don't have a sense of 'belonging' in your organization. Gallup research shows that people feel better about a job when they feel a connection to the mission and purpose of their company. It's more difficult to add great value to an organization – or feel valued or "part of the family' – if you aren't moved by its mission.
- **Do I have meaningful goals for this job?** Can you imagine still doing your job 5,10 or 15 years from now? If your answer is "Never!" then you're wasting your time, today and everyday. You will suffer gravely from the law of diminishing returns. Every hour, day, week or month you spend doing a job you hate, you get less and less benefit back. You aren't likely to be known as the best performer, best employee, or the best co-worker. You do yourself, the organization and your employer irreparable damage

everyday you show up for a job you hate. May be you're in the wrong field. May be it's time you find a new one.

12.3.10 The Change in Dimensions of Workforce

In the wake of the new economic policies throughout the world – with special emphasis on India – simultaneous pursuit of productivity and other competencies, in the industrial sector particularly, to squarely face the global competition for supreme quality goods and services and the high standards expected thereof, requires not only rigid control of work processes but treatment on the workforce or job personnel. These are assets to be developed in order to add continuing and enhanced values rather than as costs to be reduced in spite of the fact that 'automation' has supplanted/scored over the human partner and new and more flexible technologies and organizational structures are replacing our older manufacturing skills and practices. Obviously because, employees capable of improving quality require a solid grounding in the hard competencies and job knowledge but in the softer skills as well. It is common knowledge that well-prepared people too can do shoddy (bad) work or allow shoddy work to go on around them unless it is quickly realized that quality ultimately depends on the way people can effectively combine their job knowledge and technical competency with theoretical understanding of the basics of their jobs which helps them to interact with another.

Necessarily, the key to high quality achievement are personal management skills such as the ability to achieve self-esteem by setting personal goals and motivating oneself, as well as skills for influencing, communicating with and working with others upstream and downstream in the production processes. Therefore, the advances of new technologies do not necessarily represent a growing dominance of machines/robots at work. Substantial evidence to the contrary shows a growing preponderance (number) of high-skill jobs in the economy as a whole as well as increasing skill requirements on the part of the workforce in existing jobs – this means a commensurate (proportionate) increase in the quality of both technical and human elements. Such skills associated with technical complexity of work are hard bits of knowledge and physical movements necessary to perform specific tasks.

These skills include cognitive information such as the carpenters knowledge of fractions and the hand-to-eye coordination necessary for sawing straight lines. The more the knowledge and physical talent are applied, the more they are transformed from pure cognition to know how or expertise. In modern manufacturing, technicians need deep reading and writing skills in order to learn and communicate in an environment where the pace of change is faster, products are more varied and there is a premier on speedy innovation and response times. Information/technology has immensely guided in reducing repetitive thought processes or branching logic to software. He also needs mathematical skills in order to work with flexible technologies whose operations are based on arithmetic and branching logic as well as to monitor quality of output using mathematically based readouts on computers. Each employer's technologies, human-machine combinations and products are unique coming out from a broad array of specialist jobs. Thus, mechanical changes resulting in our present-day work processes and procedures require not only basic understanding of jobs and of course constant updating of employer-specific, job-oriented technical knowledge as well as learning and information essential to interact in the team. The proliferation of computer-based technologies and artificial intelligence have now compelled perfect understanding of in-house software.

Depth and breadth are not the only dimensions of skills that are charging. Skills in the new emerging economy are increasingly peripheral to hands-on work. Moreover, the context for using skill is shifting from repetitive applications to more sporadic (irregular) and exceptional uses. In addition, the content of skill requirements in shifting from job-specific to more general capabilities, from 'harder' concrete skills to 'softer' more abstract skills and from objective capabilities to more personal skills.

12.3.11 The Paradigm Shift in the Nature of Skills of Workforce

Based on surveys carried out by **Japan, USA, Australia, etc. Prof Ashok Chandra,** the then **Special Secretary, Ministry of H.R.D., Government of India** has summarized the following requirements of workers in the application of Modern Technologies like Agile Manufacturing, Bioengineering, Nanotechnology, etc. i.e., the new workers need:

(i) Multiple rather than single skill;
(ii) Breadth of skills rather than depth of skills;
(iii) Supervisory and maintenance skills rather than simple operational skills;
(iv) Behavioural skills of cooperation, coordination and group working;
(v) Cognitive skills of learning, abstract reasoning and articulation;
(vi) Flexibility and adaptability and positive attitude to change;
(vii) Diagnostic and design skills;
(viii) Capability to receive and pursue the common cause of the organization and show initiative and creativity;
(ix) Skills of learning, discovery, analysis, problem-solving, experimentation and improvement (Kaizen);
(x) Skills of planning and organizing work and linking these to the overall processes of the organizations; and
(xi) Ability to take initiative in his own skill upgradation and career growth.

13

Agile Manufacturing: Offers Many Outputs

13.1 STRATEGIC ISSUES

It is clear that agile manufacturing requires our research community to make a strategic shift away from the development of automation technology, especially in the area of decision-making, towards using technology to support people. The emphasis needs to be one developing skill and knowledge enhancing technologies, rather than skill substituting technologies. We should develop technology which leaves room for the exercise of people's skills, knowledge and judgement. Technology should be used to make these distinctive attributes of people more productive and effective. Our efforts to reproduce these unique characteristics in computers is likely to become more and more difficult as the complexity of the decision-making environment continues to increase. Most of the easy problems in decision-making have now been dealt with. The ones that are left are the difficult ones, which, if they are to be tackled successfully, will require a completely different approach.

There is need to develop interdisciplinary manufacturing system engineers; and also to develop design team structures to support interdisciplinary design. University undergraduates course need to be revised and training provided for professors as well as engineers working in industries.

Management competence also needs to be improved. Managers need to be helped to develop a stronger people orientation, and they need to become more familiar with modern management science theory and practice. Training is also needed to help managers deal with change management and to develop styles of management in keeping with the new emphasis on trust, empowerment and teamworking including:

- Education and training for lifetime of learning
- Developing social, teaming and communications skills for group working
- Learning by doing
- Increasing links between industry and educational establishment
- Open learning approaches
- A psychology of learning driven approach to the development of multimedia learning technologies
- Training of small to medium size companies and
- Broadening of skills and knowledge

Research needs to be undertaken on a wide range of topics. Consideration should be given to the following:

- The concept of agile manufacturing enterprise design. There is a need for methods and tools to support an interdisciplinary approach, with special emphasis on combining top-down and bottom-up methods, holistic approaches, organizational simulation, appropriate and selective use of technology, increased user involvement and rapid prototyping. These are all relevant tools which will be required to support a spiral approach to the design of agile manufacturing enterprise.
- More emphasis on developing decision support systems that expand the range of possible decision alternatives. Avoidance of a single minded emphasis on the spread with which decision is made with more emphasis on the quality of the decision.
- The need to address research in manufacturing as a whole, not just research into new manufacturing technologies. The need to develop new management practices, technology deployment techniques, new reward systems, more sophisticated financial justification techniques, new management accounting methods, and so on to support agile manufacturing.

Development of skill and knowledge enhancing technologies focusing on applying insight from the organizational and psychology sciences is needed. Condition monitoring, database management and networking, applications of genetic algorithms, neural networks, and so on to support decision making, vision systems, workshop-oriented programming of robots and machine tools and computer supported cooperative working are also needed.

The manufacturing industry is entering an era of competition that is unprecedented. Just a few years ago, the automotive industry business in the **USA**, for example, was shielded by the big three—**GM**, **Ford, Chrysler,** and some **foreign** producers. The quality of their products was not so high. Today, the picture has changed drastically. There are now about ten auto manufacturers in the **USA** and quality of each vehicle has improved significantly. Altogether, there are over thirty brands of vehicles available to the general public. There are many foreign and transplant operations that have cropped up recently. Manufacturing is becoming more and more global every day. The manufacturers are designing products at one place and building them at another place. The "economy of labour" the so-called fourth reach is now a big factor in the competitiveness equation. Competition is fierce in spite of the declining value of currency imbalance (that is, to dollars) and the name of the game is the value (X-ability) of the product.

X-ability = {manufacturability, maintainability, assemblability, serviceability,……}

The concept is elaborated elsewhere. In a **1987 survey: "Durability and Reliability"** and "well-made car" were the top two selling point reasons for consumers buying cars. This has changed to "quality" as the top one reason in the 1990s [Carey, 1992]. Ten years ago, these reasons ranked nowhere. Today, customers want value for their money (characterized) in some way–in a thing called "X-ability" and they will buy the product from anybody who meet their needs. This clearly shows a major shift in customer requirements. To meet customer satisfaction and to build future business, a company must not only deliver what the customers have requested, but anticipate what the customers will need next (or in the near future) and package them into the products.

- Examine and define the underlying conceptual framework on which an agile manufacturing enterprise will be built.

- Explore and understand the nature of the mass production paradigm and the nature of the culture and methodological difficulties involved in the transition to agile manufacturing.

Both in the **US and Europe**, there has been a gradual loss of competitiveness. As the **lean manufacturing paradigm** became established in **Japan**, generating competitive edge for the **Japanese companies**, the **mass production paradigm**, dominant in **US and European industry**, was contributing to this loss of competitiveness, which later became a major economic problem.

There are now a significant number of people who believe that they do not doubt lean manufacturing. But in adopting this approach we run the risk of forever chasing after a moving target, for the **Japanese** are not going to standstill and wait to be outperformed by US and European enterprises. The **Japanese** will keep innovating and perfecting their methods. Thus, adopting lean manufacturing can only be a short term, if we want to catch up with and overtake the **Japanese**, lean manufacturing is not the answer. What we need to do is something that the Japanese cannot do.

Agile manufacturing is the answer. But it is not another term for **CIM**, or any number of other fashionable buzzwords. Agile manufacturing is primarily a business concept. Its aim is quite simple to put our enterprises way out in front of our primary competitors. In agile manufacturing, the aim is to combine organization, people and technology into an integrated and coordinated whole. The agility that arises from this can be used for competitive advantage, by being able to respond rapidly to changes occurring in the market environment and through the ability to use and exploit a fundamental resource and knowledge.

- Fundamental to the exploitation of this resource is the idea of using technologies to lever people skills and knowledge. People must be brought together, in dynamic teams formed around early identified market opportunities, so that it becomes possible to lever one another's knowledge. Through these processes is sought the transformation of knowledge and ideas into new products and services, as well as improvements to existing products and services.
- The concept of agile manufacturing is built around the synthesis of a number of enterprises that each have some core skills or competencies brought to joint venturing operation, based on using each partner's facilities and resources. For this reason, these joint venture enterprises are called 'virtual corporations', because they do not own significant capital resources. This helps to make them agile, as they can be formed and changed very rapidly.
- Central to the ability to form these joint ventures is the deployment of advanced information technologies and the development of highly nimble organizational structures to support highly skilled, knowledgeable and empowered people. Agile manufacturing builds on what is good in lean manufacturing and uses what can be adapted to Western cultures, but it also adds the power of the individual and the opportunities afforded by new technologies.
- Agile manufacturing enterprises will be capable of taking advantage of the windows of opportunity that appear from time to time, in the marketplace. With agile manufacturing, will come new ways to interact with customer and suppliers. Customers will be able to gain access to manufactures' products and services and also easily assess and exploit their competencies, enabling achievement.

The key to agility however, lies in several places. An agile enterprise needs highly skilled and knowledgeable people who are flexible, motivated and responsive to change. It also needs new forms of organizational structure that engender non-hierarchical management styles and stimulate and support individuals, as well as cooperation and teamworking. Agile manufacturing enterprises also need advanced computer-based technologies.

- Agile manufacturing enterprises will bring together a wide range of knowledge in the design of a manufacturing system that encompasses suppliers and customers, and which addresses all dimensions of the system, including organization, people, technology, management accounting practices, and so on. Most importantly, the interrelated nature of all these areas needs to be recognized and an interdisciplinary manufacturing systems design method adopted as standard practice. This means going beyond the multidisciplinary approaches that are currently being adopted, and looking at areas between professions.
- Interdisciplinary design for designing agile manufacturing system depends much on a knowledge intensive era. But interdisciplinary design, however, means more than just applying knowledge from other domains, such as psychology and organizational science, to the design of agile manufacturing systems. It also implies looking into the unexplored areas between these disciplines and the areas they overlap, to find new insights, new knowledge and new and original solutions. This is one of the most important challenges that managers, system designers and integrators will face in the years ahead, for interdisciplinary design leads us to new approaches and new ways of working and of thinking. However, to successfully adopt an interdisciplinary design methods, we also need to:
    - Challenge accepted design strategies and develop new and better approaches
    - Question established and cherished beliefs and theories and develop new ones to replace those that no longer have validity
    - Consider how we can address organization, people and technology, and other issues in the design of manufacturing systems, so that we can achieve systems that are better for performance, for the environment and for the people who form a part of these systems
    - Go beyond the automation paradigm of the industrial era, to use technology in a way that makes human skill, knowledge and intelligence more effective and productive and also allows us to tap the creativity and talent of all the people involved

The challenges faced with respect to all these issues are enormous. The world of manufacturing is very complex. There is a massive number of interconnections between the various components and elements. A manufacturing enterprise is so complex that, in the past, it has been impossible to cope with it as a whole and it has been necessary to reduce it into manageable areas which have tended to be examined separately.

Tremendous problems when confronted will need us to make a transaction from traditional manufacturing to a new manufacturing paradigm such as agile manufacturing. We must face up the pain involved in making the required changes and the first step must be to appreciate the limitations of our present approaches; to understand our past failures and the nature of the changes that we will have to implement.

13.2 FAILURE THWARTS ADVANCED MANUFACTURING

The implementation of **CIM** has not resolved the problems of quality and performance to schedule as anticipated. The failure associated with investment in an advanced manufacturing technology have not been well documented. We do not advertise failure, even if it is recognized, (which sometimes is not) owing to its subjective nature. Also, our traditional management accounting systems are not well adapted to identifying failure.

There is, however, increasing evidence that the design and implementation of advanced manufacturing technologies and systems has led to significant failures and disappointments. Failures include such actions as removing equipment once installed, and many disappointments include lower than expected quality and due date improvements and productivity gains (Browne, et al. 1988; Browne, 1992; Majchrzak and Gasser, 1992).

The result of the study undertaken by the **MIT Commission** on Industrial Productivity (Dertouzos et al., 1989) points to several serious problems and weaknesses in the US manufacturing industry. Several key problem areas were identified by the Commission, They are:

- Outdated business strategies
- Short investment time horizons
- Neglect of human resources
- Failures of cooperation
- Government and industry at cross-purposes

One study of a number of manufacturing enterprises (Voss, 1988) shows that all the firms examined the achievement of technical success. In other words, all the companies got their technical systems installed and fully operational, with all operating problems resolved. However, only some of these companies managed to achieve increase in productivity and some others realized other benefits such as improved flexibility, quality, throughput times and due date performance. Amazingly, only a small number of these companies improved their competitive performance.

Several major reasons have been identified for our failures. They include:

- Technology-oriented rather than business and market-focused investment plans
- Inappropriate or non-existent business or manufacturing strategies or both
- Failure to link business and manufacturing strategies
- Inadequate assessment of the costs and benefits of investments in new technologies
- Failure to understand customer requirements and to design and build products that customers will buy
- Neglect of organization and people issues

To succeed in the modern competitive environment is an enormous challenge. Many factors, for example, rapid technological progress, environmental constraints and concurrencies are issues over which we have little or no control over, but we must live with them. Others, such as the way we design products, are more controllable and these aspects can be shaped to improve our competitive performance.

There are, however, other significant factors: for example, our traditions, which can make success difficult and drive companies in the wrong directions. Tradition manifests itself in the form of our management accounting methods, the manufacturing goals we pursue, the way we use technology to achieve our goals, our design and management practices, the organizational structures we implement and our attitudes towards people. We often ignore this issue of tradition or we only give it superficial attention. In many cases, however, our traditions are so antiquated and inappropriate that they constitute a competitive liability.

It is believed that many of the causes of our failures listed above can be traced to the question of tradition. For example, it has been our tradition to focus attention on technology to improve competitive positions (usually through cost reduction). We have often taken the organization for granted and accepted it as given. And we have ignored or at best treated as secondary, peoples issues.

We have tended to justify the CAD systems on the basis of labour cost savings in the drawings office, because this is our traditional way of justifying investments in new technology. We have justified investments in CAD on this basis, even though the benefits extend beyond the drawing office and cover issues such as quality.

13.3 TOWARDS A BALANCED APPROACH

Manufacturing in the 21st century will not be based on the traditional concepts of high levels of vertical and horizontal division of labour, standardization, and so on. Our new manufacturing paradigm agile manufacturing will not be dominated by technology, but by consideration of organization, people and other issues, in parallel with technological issues.

It is our assertion that the emergence of agile manufacturing as a new manufacturing paradigm has revolutionary implications for many of our traditional beliefs and practices.

We will also examine many of the major changes that will have to be brought about if this new paradigm is to become a reality. We will also consider some insights from the theory of change and change management and these will be used to analyze the present situation and to highlight some problematic areas.

14

Customization

The first such issue is deciding whether to create products that can be customized for individual customers or standard products that can be adapted for or by customers to meet their specific needs. Adaptive products come in two categories, customizable and customizing. With customizable products, the customer (or perhaps an intermediary), selects from various options that are present within the product the one that mostly meet his needs at that time. Perhaps the simplest customizable product to think about is the graphic equalizer, a stereo component that allows customers to select the precise among different frequency bands they desire to hear. And perhaps the most common customizable product today is the personal computer whose applications over the past decade have become highly customizable, with tool bars, preference menus, power modes and other increasingly sophisticated ways of enabling users to customize the software to their linking.

With customizing products, it is the product itself that adapts to the user. This can range from something as simple as the **Gillette Sensor razor** that, (thanks to a system of flexible springs), automatically adjusts to the contours of your face (as the advertising attests) to something as complicated as a **Matsushita washing machine** that has optical sensors, weight sensors and fuzzy logic to determine which of 600 cycles would best wash each particular brand of clothes.

In either case, because one product can be adapted to a large number of users—and can be further customized to each use—this causes no extraordinary changes in manufacturing as do customized products. It is, therefore, the most comfortable path to customization taken by those still ensconced (established) within the heritage of mass production. Automakers, for example, have been constantly adding adaptive (both customizable and customizing) features that embed multiple permutations within a standard design particularly a valid development strategy because cars are often driven by various people. Recent announcements include an auto static feature on **Chrysler Eagle Visions** that allows drivers to switch between automatic and standard transmission; push-button memories on **Lincoln Continentals** to adjust seating positions for two different drivers, and furthering these innovation, a "self-adjusting comfort seat" under development by **Johnson Controls** that wraps around and adjusts to the proper pressure points of any driver.

While these are appropriate steps for mass producers, there is a real danger here. It is imperative for any development organization not to blindly latch on to adaptive products as the solution for all time. This would foreclose many facets of customization that are or will become very important to customers.

The most feasible approach is generally to provide some degree of both customization and adaptability, particularly for sophisticated or complex products with a high degree of functionality. One company that does a superb job of managing both of these facets is lighting-control manufacturer **Lutron Electronics Co, Inc. of Coopersburg, Pennsylvania**. Many years ago **Lutron** was the first in its industry to embed microprocessors in its products to give people great control over their own lighting environment. Customers can achieve maximum productivity (in offices) or design appropriate moods (in homes) at different times for different occasions. The design tool must manage the complexity, that, on the one hand, allows you to tailor your products to individual customers, while on the other threatens to choke your customers' ability to quickly, easily and without hassle find exactly what he wants. For example, **Anderson Corp's Window of knowledge** is a very sophisticated design tool provided to distributors to enable their collaboration with home owners on custom window design. Its success is predicated on highly trained personnel that know windows, know customer needs and know how to use the design tool to guide customers to the right solution.

Less complex products, however, need less complex solutions, such as that provided by writer sheet music. It digitizes the music nations of each song and places them in disk storage within an electronic **touch-screen kiosk**. Rather than browsing through hundreds of different choices by hand and several thousands in larger stores, customers simply walk up to the kiosk and start typing in the title or the composer. The system automatically presents and narrows their choices as they type until exactly the right choice is found. It can then play back and display the available versions, transpose the song to any key desired and immediately print each customer's exact choice.

14.1 THE FORMIDABLE CHALLENGES OF CUSTOMIZATION

The challenges in mass customization are not just offering the customization, per se. Craft industries did that for centuries before the Industrial Revolution. Rather, the challenges are to:

- Build customized products quickly. Being able to order quickly in a flexible manufacturing environment will accomplish this.
- Build customized products at low cost, that is to show how mass customized products can achieve low cost by design.

Cost must be designed out of the product and production processes, as it is very difficult to remove cost through 'cost-reduction' measures after the product has been designed. Cost 'reduction' rarely even pays off the expense of the cost-reduction effort.

The key to achieve the lowest-cost product is to base all thinking and decisions on a total cost perspective. Unfortunately, the typical company cost system reports only material and labour costs. All other costs are called overheads, which are spread over all corporate activities according to some arbitrary allocation algorithm, such as some portion to material, labour or processing cost. And yet, all products do not have the same overhead demands. In fact, much can be done to lower overhead costs by design.

There are two categories of cost: namely, "reported costs" such as labour and materials and "unreported costs" such as overhead costs. Advanced design techniques can be used to design low-cost products for the minimum labour and materials costs and then build them very efficiently for the minimum overhead cost.This has two goals.

- One goal is to suggest several categories of cost savings over more conventional operations. Using total cost methodologies, this cost saving can be calculated. This will be especially important if mass customized products must share production facilities and some of their overheads, with non-customized products. The lower overhead demands of mass customized products will need to be considered when overhead costs are allocated.

- The second goal is to show all the monetary benefits to mass customized production so that the reader will be motivated to implement and support the programme that will generate these benefits: advanced product development methodologies, parts commonality, modular design, setup reduction, just-in-time, flexible manufacturing, and build-to-order, which together yield a production organization capable of mass customization.

14.2 AIM TO ACHIEVE LOW COST

Low-cost products do not come from volume alone; that is the mass production paradigm. With proper design and flexible operations, mass customized parts can be made at low cost virtually. Low-cost products do not come from cheap parts, which are often chosen because they appear to lower the reported material costs. However, cheap parts will usually explode other costs of quality, service, operations and other overheads. Similarly, cutting corners in any manner will probably end up costing much later. Total cost accounting can help by quantifying all costs, but until it is known, product designers must rely on total cost.

An unwise strategy to reduce cost may be to omit features. Although this may reduce cost, it may also reduce revenue and hamper product evolution.

Cost is very difficult to remove after the product is designed. As shown here, 80 per cent of the cost is designed into the product and is very difficult to remove later. Changing the design to reduce costs may force other changes. Trying to significantly lower the cost after production release is usually futile because of many early 'cast-in-concrete' decisions, which limit opportunities. Finally, the total cost of effecting the change may not be paid back by the cost saving within the expected life of the product. Few companies really keep track of the total costs of changing designs.

There are also intangible impacts of excessive focus on cost reduction. It absorbs effort and talent that could be applied to more productive activities, such as developing better new products.

Low-cost products do not result from 'saving' cost by cutting product development and continuous improvement efforts. Low-cost products do not result from manufacturing misconception like purchasing policies based on low-bidder or offshore manufacturing to lower labour costs, which too often raise many other overhead costs and lengthen delivery times.

Nearly 80 per cent of the lifetime cumulative cost of a product is determined by the product's design.

Important for mass customization is that 60 per cent of the product cost is determined by product architecture. Low-product design is based on the idea that cost is designed into the product, especially by early concept decisions.

Tools such as **DFM** can help design products that are easier, and thus less costly to build. **CE** can ensure the lowest-cost processing since the processes are concurrently designed with the product. Quality can be designed into the product with 'robust' design techniques **(Taguchi methods)**. This is built into the product with process controls instead of more expensive inspection paradigm. Maximum utilization of catalogue hardware can minimize part cost. Involvement of vendors early can result in lower-cost outsourced parts.

14.3 LASTING RELATIONSHIP WITH CUSTOMERS

In today's hotly contested competitive environment, the most worthwhile asset of any corporation is its long-lasting relationships with its customers.

A rapidly expanding array of interactive technologies including electronic kiosks, on-line services, e-mail, fax response, worldwide websites, and so forth enables companies to efficiently dialogue with thousands and potentially millions of individual customers. By interacting with them on a personal basis, companies can better learn the wants, needs and preferences of each and can then fulfil those preferences with better tailored goods and services. The more the customer teaches the company, the better it can provide exactly what he wants, when, where and how he wants it, and the more difficult it will be for him to be attracted. Even if a competitor were to build exactly the same capabilities, a customer already involved in a learning relationship with a firm would have to spend an excessive amount of time and energy teaching the competitor what the firm already knows.

14.4 FLEXIBILITY (AGILITY)

Flexibility can reduce overhead costs significantly and can put the 'mass' in mass customization. There are some interesting parallels between flexibility and quality. Twenty years ago, it was commonly believed that quality costs more. Then **Philip Crosby in his book 'Quality is Free'** showed that the gains from lowering the cost of quality would pay for quality, thus making it free. Similarly, the financial gains derived from flexible operations can more than pay for the effort to make operations flexible. These gains will become a source of competitive advantage over competitors that do not embrace mass customization.

There is also a lot of working capital tied up in a various forms of inventory: raw materials and parts inventory; work-in-process inventory; and finished goods inventory in factory warehouses, at distributors and at the dealers.

Several opportunities are open to lower the costs of inventory and other overhead costs by designing flexibility into products and plants.

As pointed out earlier, many companies offer goods but do not do it cost-effectively. By mass customizing products, the customization process is build into the system, i.e., the product design and the manufacturing operation.

The extra manufacturing cost to ad hoc customization and configuration of 'reactive' product is much more than it would be for mass customized products. These customization costs often include extra tooling, lengthy setups for small runs, inefficient production control, low equipment utilization, special programming, slow and frequent 'learning curves', special tests and inspections, and lots of 'fire drills' to shove customized products through a mass production factory. Further, these low-volume customized products may disrupt the manufacturer of the standard products, thus increasing their cost. Thus, in companies with traditional accounting systems, custom products are really subsidized by the standard products.

A large part of working capital is tied up in the cost of variety. Eliminating setup and reducing the lot size to one eliminates most of the cost of variety. True one-piece flow can eliminate inventory or reduce it to one piece between stations.

The key element to just-in-time operations and manufacturing flexibility is setup elimination. If setup can be eliminated, the operations are said to be flexible, meaning that every product could be different and still reap mass production efficiencies.

Concepts based on common parts encourage common parts usage. Floor stocks can make common parts readily available for engineers to evaluate the parts, conduct experiments and build breadboards. Floor stocks can also be mounted on display boards in prominent places near the design team.

There are also other means of achieving cost-effectiveness such as: usage of personal display boards, specification books, cost matrices, set-up elimination, inventory reduction, internal material logistics, flexible manufacture, quick deliveries to customers, build to order, quicker deliveries from vendors, cost reduction in purchase costs, etc.

14.5 MASS CUSTOMIZATION

14.5.1 New Compulsion in Business

Mass customization has been the new compulsion in business for success of the many profound changes to which businesses must respond to succeed in today's turbulent climate. The initiatives of most companies—measuring customer satisfaction, implementing quality function deployment, putting together cross-functional teams that span marketing, manufacturing and development and so forth—are in fact market-focused. Customers are not sought but it is done to create better averages rather than customized goods. The imperative today is to understand and fulfil each individual customer's increasingly diverse wants and needs while meeting the co-equal imperative for achieving low costs. The critical role of executives, managers and engineers within the product development community is harnessing new technologies, manufacturing capabilities and management techniques to efficiently serve each customer uniquely.

The task of mass customization is to do it efficiently at a price customers are willing to pay and at a cost that allows for profitable margins. Most companies today handle variety by layering new work activities on top of existing processes and handle customization through exceptions and expediters—all of which add tremendous costs, waste time and need extra resources. But mass customization is not everything to everybody; rather, it is doing only and exactly what each customer wants, when they want it. Most mass customizers find that although there may be a significant upfront investment in developing the products, markets become fragmented enough when mass production techniques can no longer effectively predict what customers desire.

Further, most customers are willing to pay a premium (often 10 to 50 per cent) simply because customized products have greater value than standardized ones—they more closely match each individual's needs. Margins, therefore, often increase greatly, especially once a company realizes the gains from eliminating the carrying costs of finished goods inventory, having customers not to walk away when they cannot get what they want and not having to put some product on sale because no one wanted it. This obviously leads to think about the financial gains in the company that could be realized through the elimination of finished goods inventory alone.

Because of these tremendous advantages. Mass customization is moving from the frontier to fast becoming an imperative in industry. It is the Next Logical Progression in Business competition. This can be seen clearly through a framework (Ref. Fig. 14.1) originally developed by two professors of the **University of North Carolina, Bart Victor and Andy Boynton, and other collaborative contributors.**

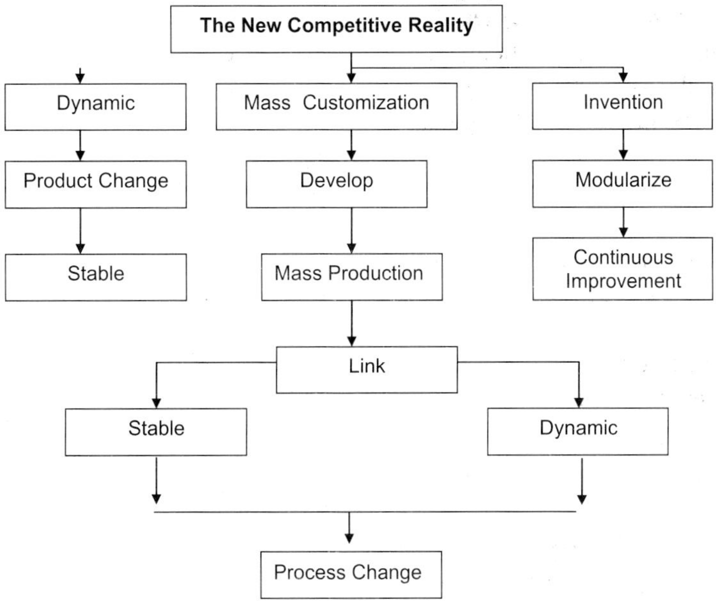

FIGURE 14.1 Next logical progression in business.

Source: Bant Victor, Andrew C. Boynton and B. Josep Pine II.

For centuries all businesses were craft producers that basically followed the invention model—if any customer wanted something, the company did not know how to make it. With the Industrial Revolution and in particular **Henry Ford's** development of the assembly line came the capability for mass production. Here, everything was stable: these organizations found the one best way to produce a given product and then move down the learning curve as fast as possible to do it. Both product and process change come only very slowly to ensure fixed costs are recouped. Every once in a while (typically four to five years or longer), mass producers had to depend upon some other invention organization (usually their own R&D laboratories) for a new product to be mass produced.

14.5.2 The New Frontier: Need for Relationship between Mass Production and Invention Organizations

Much has changed since then. Not only there are more companies mass customizing their products, but also as the example above shows, they are doing so to a breadth and depth that is often astonishing compared to what could be done just five years ago. Much has also changed in what mass customization truly is. By the 1980s, the old system of mass production was not working any more and it needed to move to a new system that began with greater and greater variety, reinforcing feedback loop with customers and blending more and more customization until every customer was able to purchase unique goods or services.

Mass producers were thus very dependent on invention organizations to create new products, and invention organizations were often dependent on mass producers to provide a ready market for their highly differentiated creations. This synergy between mass production and invention organizations worked very well for a very long time; it was at the heart of **American companies** (in particular but not exclusively); they figured out that if they continually improved their processes, they could achieve both lower costs and higher quality than the typical mass producer. By embracing dynamic process changes, they could gain a significant advantage over their competitors. This was so different from the old ways of doing things that it took **American producers** a long time to figure out exactly what it was. Now most companies subscribed to the continuous improvement business model (at least in principle, if not fully in practice), incorporate such innovations as statistical process control, cross-functional teams, total quality management and customer satisfaction measurements, to name just a few. At its ideal, organizations following this model have a process life cycle of one execution: every execution is different and better than the last.

While companies everywhere seem to be making great strides in quality by focusing on continuous improvement, many firms are already moving beyond this to the mass customization business model. Stable but very flexible and responsive process provide a dynamic flow of products, enabling the companies to achieve both low costs and high individual customization. In this business model, the organization's primary thrust is to identify and fulfil the individual wants and needs of every customer. Ideally, the product life cycle is one unit: every product is different from the last and uniquely suited to that particular customer's needs.

The organizations can move between business model, but based on the researches of Victor, Boynton, et al., as well as direct client experience, there is a definite order to the transformations that must occur to reach mass customization. The first such transformation, from invention to mass production, is called 'development' and is a well-known activity. It requires articulating and stabilizing the product and process, making it definable and repeatable for high-volume, low-cost production.

While that is what often occurs, most companies use product proliferation to forestall true customization, trying desperately to maintain the heritage of mass production in the face of rapidly fragmenting markets. Today, we can now be very clear: variety is not the same as customization. Variety is producing a product and putting it in finished goods inventory in the hope that some customer will come along and desire it. In contrast, it is customization only if its produced in response to a particular customer's desires. The term individual customization is in fact redundant. Variety is about giving more customers more choices in the hope that they can find something close to what they need. But often we overwhelm customers with so much proliferation that they throw up their hands at having to go through a lengthy decision-making process with little or no support and simply walk away. Fundamentally customers do not want choice; they just want exactly what they want. Your task is to figure out (often through collaboration) exactly what they need and then produce it.

The resulting mass production organization has historically been very hierarchical and bureaucratic, with very little information flow between functions. To enable the organization to switch to continuous improvement, these vertically separated functions must be linked through cross-functional teams, information sharing, and a horizontal process focus. The result is a set of tightly coupled, high quality process, capable of constant improvement.

To move to mass customization requires that products should be modularized to provide unique combinations for any customer. This enables companies to deliver efficiently individual modules to the customer value—whether they be a particular length of pant, a specific pitch of a window, a certain pH factor in industrial soap or an exacting pneumatic valve connection within the structure of a total product architecture.

Further, the tightly coupled processes created through continuous improvement should also be broken apart and modularized, so that, at its ideal, specific process modules can dynamically link to other process modules which are required to create the unique end-to-end value chain that will best satisfy each individual customer. This results in a dynamic network where linkages between people and processes are loosely coupled to enable the mass customization of goods and services.

Mass customization organizations have also to be able to renew themselves, to realize when they cannot satisfy a particular customer or cannot go after a particular market opportunity with their current envelope of customization. This causes them to reinvent a new product module or a new process or to link themselves to an organization inside or outside the firm to provide the new capability required. And at times, firms may have to overthrow their entire product or process architecture—before their competitors do—and reinvent one that will once again provide a distinct competitive advantage.

Although the organization will never want to "live" within mass production again, it does have to go through the activities outlined above for each new product, process, or organization module created in invention. Each one still has to be developed and stabilized, linked to the rest of the organization and be made of high quality and finally, it should be made to fit the modular architecture for mass customization.

There are so many degrees of freedom for what, how, where and when a company customizes that it can open a large amount of competitive space. And much of that freedom depends upon what occurs in product development. For any company to realize the full potential of this business model, those responsible for bringing new products to market must fully embrace it as an imperative as well. Without the support, diligence and effort required to create products that can be mass customized, no company will fare well.

There are a number of issues concerning how best to do this for any particular company, ranging from the nature of the product to how it is represented to customers; from how to create an infrastructure to produce it to how to determine when to renew it. These crucial issues are discussed below.

14.5.3 Agile SMLEs (Small, Medium and Large Enterprises) for Mass Customization

Most manufacturers are being forced to offer more variety to satisfy customers and compete in niche markets. Many also need to customize products for increasingly selective customers. However, if products are not designed well for this and manufacturing is not flexible enough, then offering more variety of customizing products will be a slow and costly ordeal.

Agile Product Development for Mass Customization illustrates how you can:

- Reduce costs by understanding, quantifying, and minimizing the "cost of variety."
- Rationalize product lines to free operations from low-margin products. This can significantly improve profits, release valuable resources, improve flexibility and encourage the development of mass customized products.
- Lower costs and improve flexibility, by standardizing parts, processes, features, tools, and raw materials [as a prerequisite for JIT, agile manufacturing, and mass customization or as a stand-alone programme]
- Put the mass in mass customization and create custom products cost-effectively.
- Implement agile manufacturing and build products quickly to order without setup or inventory.
- Implement an agile product development process for ultra-fast time-to-market, where "new" products are actually planned variations on a theme.
- Concurrently engineer product families and flexible processes and design agile products around modular architectures, common parts, versatile modules, standardized interfaces and standard processes.

Through these leading-edge techniques, Agile Product Development for Mass Customization will allow your company to enhance agility, improve variety, reduce costs, develop products more quickly, speed delivery and satisfy unique customer needs.

Therefore, the structural and organizational requirements of agile enterprise should include important leading edges like:

- A flat, fast, flexible organization, with continuous interaction, support, and communications among various disciplines, with no fat in the middle
- Highly decentralized management that recognizes what knowledge base is and how it can manage that base most effectively
- A multi-skilled, flexible labour force in which blue-collar workers will more closely resemble white-collar workers in activity and responsibility
- Extremely flexible strategies and relationships with suppliers and even with direct competitors (these arrangements allow partnerships to be formed rapidly both within and outside the core sphere of expertise or business and disbanded when new opportunities arise; i.e., "virtual" enterprises)
- Complete mastery of lean production and concurrent engineering
- Flexible tooling and automation
- A deep understanding of the customer based on operating integration and real-time market data collection and analysis

- And, most important from the perspective of this outlook, a complete mastery of the tools of continuous transitioning.

With markets becoming more and more dynamic and competition increasing based on improved quality and customer responsiveness, the development of an appropriate manufacturing strategy is likely to have a major impact on our survival, success and growth. Therefore, in general, the strategy for agile manufacturing should be focused on designing an enterprise that will lead to :

- Faster response to highly variable customer demand patterns
- Improved productivity
- Opportunities for system wide innovation, learning and improvement
- Improved product quality
- Better utilization of expensive capital and improved return on investment
- Improved customer and market focus and a better understanding of customer needs and closer customer relationships
- Flexibility to cope with a wide range of batch sizes including one-of-a-kind production, and a wide range of products
- Integration of suppliers into product development and manufacturing processes
- Short-order capability and the ability to rapidly respond to new windows of opportunity
- Capability to undertake multi-venturing through virtual corporations
- Reduced operating expenses through reduced stock holding, less scrap and re-work and reduced system down-time
- Shorter design and manufacturing lead times
- Reduced indirect labour and other overhead costs
- More time and opportunities for management to tackle problems
- System integrity and robustness

These goals provide the basis for agile manufacturing enterprise strategy. But, these goals will only be accomplished by investing in proper organization, people and technology. A word of caution is necessary at this stage, it is important not to confuse agile manufacturing with flexible manufacturing, lean manufacturing or computer integrated manufacturing. What then does agile manufacturing mean?

Agile manufacturing requires the integration of organizations, people and technology as a coordinated whole. This is what delivers agility. Agility is not a function of one single dimension. As such, it must be multi-skilled having a flexible team of working people who are empowered in the total enterprise and have thorough knowledge of management skills and enhancing technologies to achieve continuous improvement while carrying out the change and risk management.

It should be stressed that these generic features are neither a prescriptive framework for agile manufacturing, nor are they intended to be a prescriptive solution to the various problems that beset manufacturing enterprises. The needs of manufacturing companies are not generally amendable to prescriptive solutions. All in all, the above explained working environment to succeed in agile manufacture demands a sustainable situation all through where an 'Information Network' indeed an 'Internet Solution' will be an indispensable aid.

14.5.4 Network for Success

This is in tune with the Indian industrial and global success which demands a radical repositioning of Indian business using the Internet, web tools and a global market approach. The success of this in turn depends on the integration of InfoTech Software's India, Business Planning, Innovation, Government Policy and Entrepreneurial orientation. Multiplication of the success will become further possible through networking; sharing of experiences, relevant to visualizing business ideas, understand the trends of accelerating technology, acquisition of knowledge, generation of cooperation, financial partnership and building of virtual institutions. As such the aim should be to have a forum for reaction with venture capitalists, business planners consultants, technologists, builders, government institutions, world bodies, etc., as well as providers of cutting-edge technologies, educational research institutions, etc. To improve the quality and the context of the restrictions imposed by escalating cost, low flexibility and limited reach there is positive need to change our strategies and march with the advent of the internet.

14.5.5 Comparison between Lean and Agile Manufacturing Methods

The comparison between lean and agile technologies on manufacturing is presented in Fig. 14.2.

14.5.6 Comparison between Mass Production and Customization

The comparison is shown in Table 14.1:

TABLE 14.1

Mass Production	*Mass Customization*
(i) Factories become geared up to produce large volume of low variety and low cost production, but they become inflexible and lose the capability to respond to rapid shift in market conditions.	(i) Factories become geared up to produce one or one million of high varieties and low cost production, but they become highly flexible and gain the capability to respond to rapid shift in market conditions.

Contd.

Mass Production	*Mass Customization*
(ii) There is little chance to enhance competitiveness achieving less time to design, less inventory, less dynamic workers, less identification with market opportunities, less improvement and lower quality in product, less highly skilled knowledge and empowered people.	(ii) There is maximum chance to enhance competitiveness achieving with minimum time to design, no inventory, no defects, no exploitation of resources, highly dynamic workers, high identification with market opportunities, maximum improvement and higher quality in products, better highly skilled knowledgeable and empowered people.
(iii) To remain competitive and profitable in a price-driven market environment, we have to continually reduce costs and traditionally achieve this objective through mass production (economies of scale), product rationalization, division of labour and application of technology to achieve increased capacity, improved machine utilization, automated production and the de-skilling and elimination of direct labour.	(iii) Integration of Management, Workforce and Technology into a coordinate, independent system which is the competitive weapon to exploit agility to achieve competitive advantages.
(iv) Increasing demands by customers for better quality products, greater choice in relation to more customized products coupled with rapid technological developments resulting in pursuit of manufacturing and product innovation, continuous revision and updating of products lead to a drive to reduce inventory levels, improve quality and increase flexibility. These are not so easily possible in **mass production**. The management accounting methods employed in labour-intensive mass production are no longer applicable today.	(iv) All the demands of mass production are squarely met with by its structure for competent global competitiveness.
(v) Not at all sustainable	(v) Sustainability is achieved.
(vi) Concurrent engineering is not involved.	(vi) Concurrent engineering is involved.
(vii) No need of market survey and concur to the specification of customers, thus no delay in production.	(vii) Need market survey and concur with specifications of the customers and thus delay in production.

14.5.7 Road Map on Agile Manufacturing

Agile manufacturing, has been already described as the science of business system integrating management, technology and workforce, making the system flexible enough for a manufacturer to switch over from a production component/product to another component/product in a cost-effective manner within the frame work of the system. This term has been first introduced with the publication of the Lacocca institute report "21st Century Manufacturing Enterprise Strategy". The impact of these three aspects on manufacturing is shown as a Road Map of Agile Manufacturing in Fig. 14.3.

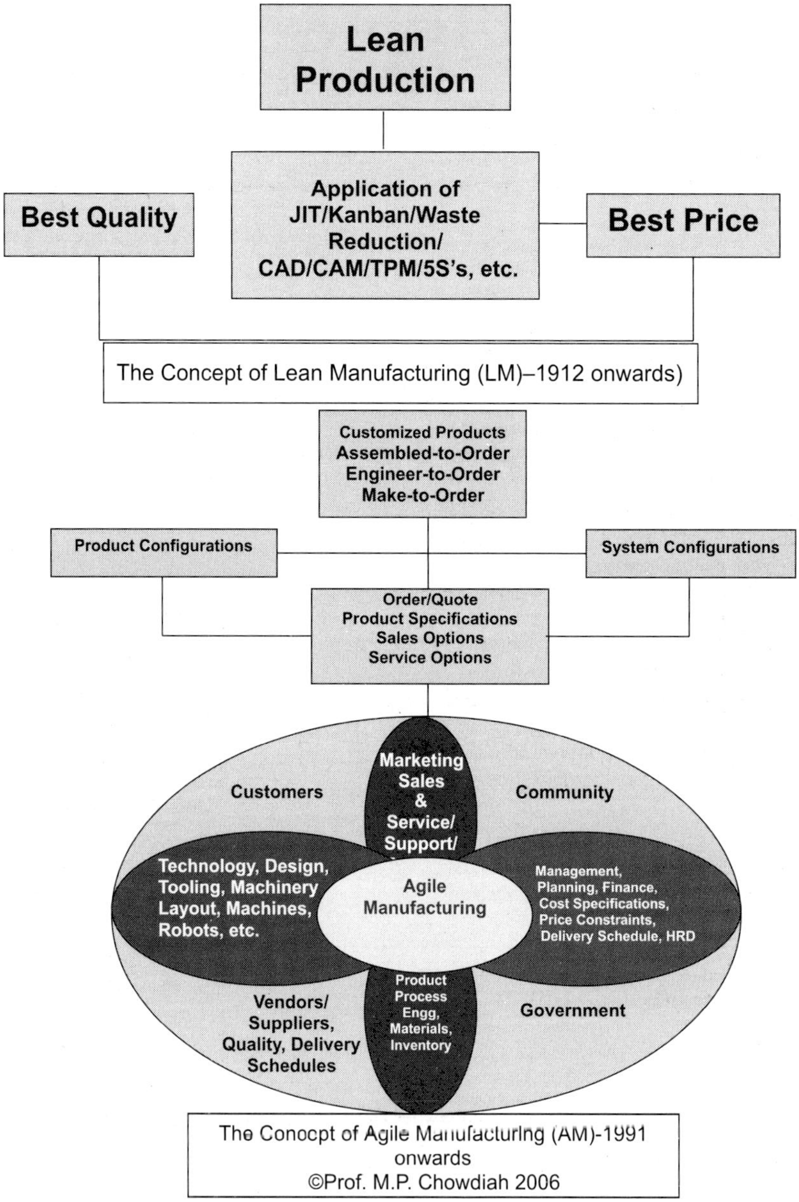

FIGURE 14.2: Lean and agile manufacturing methods.

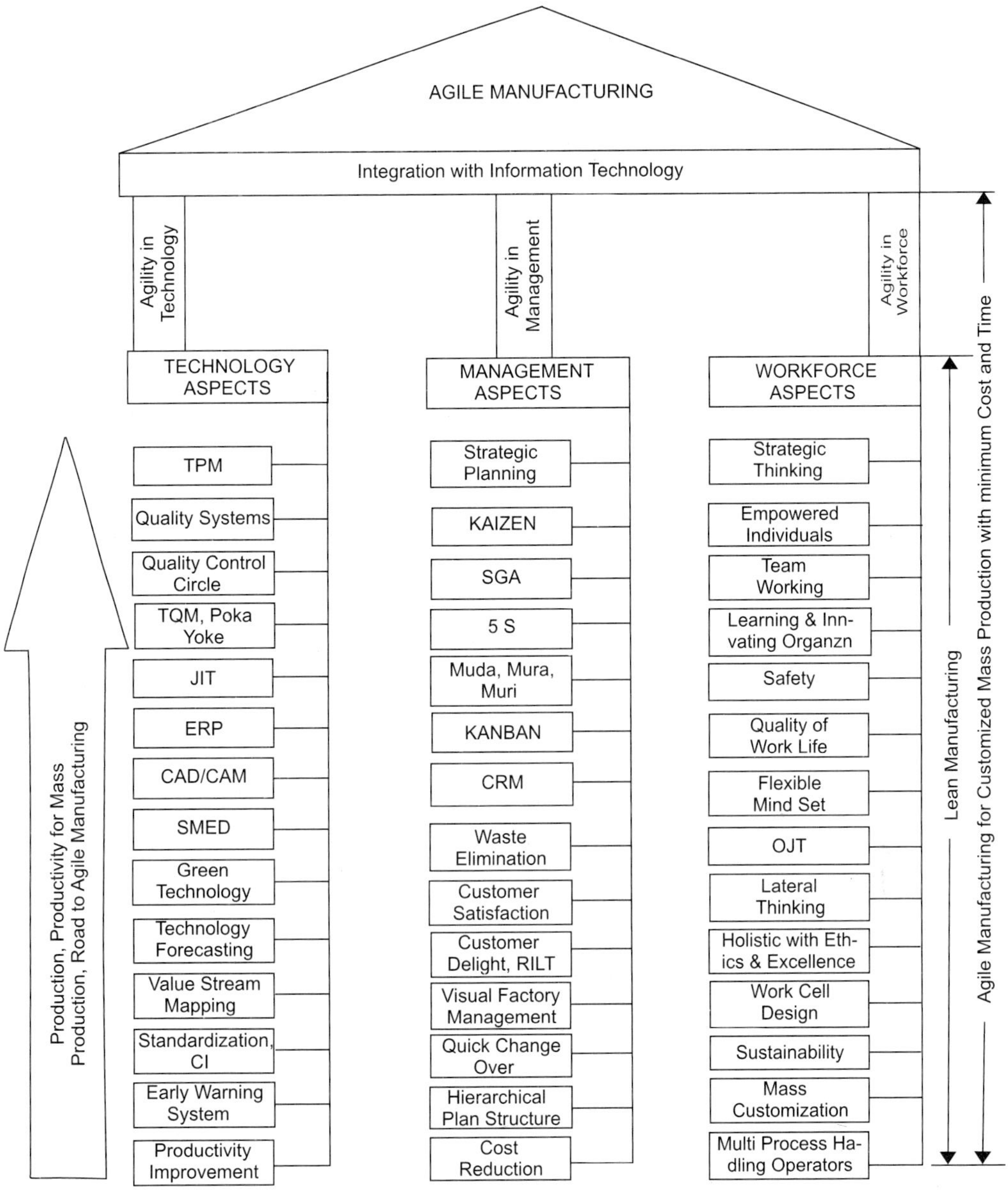

LEGEND:
TPM = Total Productive Maintenance
TQM = Total Quality Management
JIT = Just-in-Time
ERP = Enterprise Resource Planning
CAD/CAM = Computer Aided Design/Computer Aided Manufacturing
SMED = Single Minute Exchange of Dies
CI = Continuous Improvement
SGA = Small Group Activity
CRM = Customer Relations Management
RILT = Reduction in Lead Time
OJT = On the Job Training

FIGURE 14.3: Road map to agile manufacturing.

15

Japan's Just-in-Time Production (JIT)

This topic has already been covered earlier in various bits but more of it is given here. Unfortunately, an element of complacency and arrogance crept into many **American companies**, and the strides being made by **European** and more particularly **Japanese companies** were recognized. During the 1970s foreign competition became a major factor in the **United States**. Entire industries (consumer electronics, motorcycles), photographic equipment, for example, were wrested out of the hands of major **US Corporations** by vital and innovative **Foreign companies**. The **US** automotive industry was impacted significantly by the influx of Japanese products that were better made, better designed, more in touch with the market, and less expensive. The automobile giants that had dominated not only the US car market but also much of the international market were left reeling as their market share and profitability fell dramatically.

Many theories have been proposed to account for the success of Japanese companies over the last 25 years and several factors have contributed to their success. These factors include a government-planned economic development programme with low-rate long-term loans, low-wage workforce that was prepared to sacrifice in order that companies may prosper and an awareness of designing to meet customer needs. One crucial aspect of Japanese success in the Western marketplace has been innovative new production techniques. These techniques include an emphasis on quality, JIT production control, employee involvement and production flexibility. These techniques have been given a number of different names and even more acronyms, but agile manufacturing best describes the series of techniques and approaches used by good Japanese companies to provide customers with high-quality, well-designed, reliable products that have swept the world markets in diverse industries from ship-building to consumer electronics.

A sad aspect of this scene is the fact that the majority of the ideas and techniques which made these **Japanese corporations** successful had ironically been developed in the **United States**. The crucial use of innovative production and quality-control techniques was taught to Japanese companies by people like **Dr W Edwards Deming** and **Joseph M Juran in the 1950s. US Corporations** at that time did not acknowledge or foresee the importance of a TQC approach to their manufacturing business. The concept of **JIT** manufacturing to a large extent pioneered by **Henry Ford in the 1920s,** had not been carried through into post-war production technology.

While lacking clear or formal definitions, the terms 'agile manufacturing' is one of those irritating pieces of jargon that has grown up around manufacturing industry of late. However, the term is useful and does capture the importance of changes currently taking place in the manufacturing industry. To be able to compete in world markets requires for more than just implementing a few new techniques. Companies must transform operations so that their products, quality, and service are second to none. Agile manufacturing is not a set of techniques—it represents a total change in the way manufacturers approach their business.

15.1 CHANGES IN MANUFACTURING INDUSTRY

The concept of manufacturing industry is relatively new. Prior to the industrial revolution, goods were manufactured by craftsmen in small private workshops employing either an individual or a small group of skilled people making products one at a time. Things were made that way until the industrial revolution brought large numbers of works off the farms and into the burgeoning cities to labour in factories. This revolution occurred in **Europe** during the eighteenth century and a little later in the **United States**. These changes were brought about by the brilliant thinking of a few men who developed the concepts of making products from interchangeable parts and who invented the machines and processes of mass production.

All these changes revolutionized the **western society** by changing the nature of work, changing the demographic landscape of countries, altering family life by creating job external to the family home and farm, and by creating unheard-of prosperity. There were also significant negative repercussions. Abject poverty and degradation of people in the industrial cities was vividly portrayed by **Charles Dickens** in novels like **Oliver Twist. Fredrick Engels' descriptions of living conditions in working-class Liverpool, England,** in the early nineteenth century shocked the nation and spurred on his friend **Karl Marx**, who was at the time developing entirely new economic theories. Their community manifesto called for the workers of the world to unite a 'throw off your chains' and inspired the likes of **Leon Trotsky and Mao.**

15.2 NEW MANAGEMENT METHODS

Throughout the nineteenth century, as industrial development became more and more sophisticated, new organizational methods were required to manage the emerging companies.

This change fostered the ideal of professional management of industrial organizations. Railroads, the telegraph (and later the telephone), and improved roads opened the possibility of diverse organizations controlled from different locations. Brilliant new techniques were developed by companies like **Sears Rebuck, Dupont,** the steel industry and railroads to maintain control and productivity with these more complex corporations.

The **'Scientific Management' movement**, developed in the late nineteenth century, emphasized the time study of industrial processes on the 'right' way to do any task to achieve maximum productivity. The introduction of mass production lines by **Henry Ford** revolutionized production technology and the conglomeration of smaller companies into a single central giant corporation (like **General Motors** in the **United States of Imperial Chemical Industries in Britain**) revolutionized industrial management. These concepts of professional management were legitimized in later 1920s and 1930s when the great universities and learning centres such as Harvard developed business schools to teach these dynamic new techniques. The introduction of professional associations followed to codify and regulated these new approaches.

The challenge for all Western manufacturing companies at present is to implement the aspects of agile manufacturing that will enable them to compete in the world market. It has become eminently clear that no industry is exempt from the onslaught of competitive pressure from companies (usually foreign companies) that are able to present better service, quality, design and flexibility.

There is a room or some optimism here. Many Western companies have recognized the competitive challenge and have been able to make the radical changes required to meet and beat that challenge. Some of these companies are well known, large, multinational corporations and others are small, privately held businesses. The size and complexity was of little relevance to a company's ability to innovate and succeed in the turbulent tides of 1990s manufacturing. In some ways, larger corporations had a more difficult task than smaller organizations because they had to cut through years of entrenched bureaucracy in oraer to make the changes needed. Smaller companies were often quicker and better able to respond to competitive challenge.

15.3 VITALITY OF AGILE MANUFACTURING AND ITS IMMENSE POTENTIAL

Agile manufacturing applies a series of techniques employed by good companies to bring about unprecedented improvements in quality, productivity and customer service. These techniques are not new, many have been available for several decades, and others have been developed gradually over the last 30 years by innovative companies like **Toyota**. What is new about these techniques is that, when they are implemented in products, their responsiveness to customers, innovation and flexibility is clear and emphatic. These improvements give companies such an edge in the marketplace that they are able to take a commanding competitive lead.

Agile manufacturing does not so much represent a series of techniques as a fundamental change in management philosophy.

The sequence for success in agile manufacture is significant:

- Stress on quality
- Just-in-time manufacturing
- People
- Flexibility

This list represents the amount of interest and involvement observed among **Western manufacturers** in recent years. Most companies have got the message on quality. Everyone knows that to compete in world markets requires levels of quality that far exceed previous expectations and most companies have some kind of active quality improvement programme in place.

Many companies have embraced at least some aspects of **JIT** manufacturing, and some have a fully developed **JIT** implementation programme addressing the issues of cycle times, batch sizes, vendor relations and synchronized manufacturing.

Changes in the way people are managed within the organization is very important. Such changes are difficult because the issues deal with deeply held cultural understandings, and several years of consistent effort is required to make changes of this sort.

Very few Western companies are serious in addressing flexibility as a weapon in the battle for world markets. There are some notable exceptions, but the average Western manufacturer seems unaware of the issues.

In agile manufacturing, quality is a way of life. The approach taken to quality is quite different from that of traditional manufacturers because agile manufacturing is concerned not only with detecting quality problems but also solving the source of those problems. A traditional manufacturer, recognizing that there will always be rejects with any manufactured item, will often build an acceptable reject level into production plants and customer orders. An agile manufacturer recognizes no such 'reality' and sets the quality go for zero defects or 100 per cent quality. This level of quality must be achieved by a systematic, long-term programme of identifying quality problems and harnessing the entire workforce resolving those problems.

A traditional manufacturer employs a large staff inspector whose job is to check the quality from all materials purchased from vendors and everything manufactured within the plant. Subassemblies are inspected at each stage in the production process to ensure that they 'pass'. The thinking behind this approach has always been to ensure quality by having the product inspected carefully at each stage in the production process by independent, trained inspectors rather than by the operators making the goods.

This approach is not only expensive but also ineffective. Many companies have found that improved quality does not come from having more inspectors. The production state does not feel that product quality is its responsibility if its work is inspected by others throughout the process. The inspection department is considered to be responsible for product quality. This situation is untenable for all concerned inspectors thoroughly frustrated since they are held responsible for something over which they have no control. An inspector can do nothing to improve product quality; all he or she can do is to find faults after the event. Production operators are equally frustrated because they have people looking over their shoulders all the time checking up on them and assigning blame when things go wrong.

An agile manufacturer places the responsibility for quality on to the people who do the job. A pride of ownership is fostered on the shop floor because people have both the authority and responsibility for activity. Craftsmen have to represent themselves. Much of this outlook has been lost, as modern industry has developed from craft-based to factory-based production. Agile manufacturers can engender this kind of pride and personal responsibility within the production workforce and beyond. For this approach to be successful, people responsible for quality must have the tools, training, and authority to ensure zero rejects. A number of techniques that lend themselves to this approach are widely used. These techniques include **SPC**, the posting of quality results on the shop floor as they occur, quality circles where quality problems can be analyzed and resolved and appropriate education and quality standards that emphasize the importance of quality to the company.

The quest for quality does not stop at the shop floor. Every department of the company is affected by the need for zero defects. Because the design of the product often has more bearing on quality than the production process, many agile manufacturers change the way products are designed to incorporate the concepts of quality. The change requires a much closer link between design engineers and the people who make the products. Some companies, including some of the very best **Japanese companies** have developed a radically different design approach. People from all aspects of the business are brought together into a team to work with design engineers in the development of new products. The purpose of these teams is to design the products the customers want, to produce products at a price they are willing to pay, to achieve 100 per cent quality, and to bring innovative ideas to market quickly. This ambitious objective has been highly successful. Sales people, marketing people, design engineers, production engineers, cost accountants, field-service people, shop-floor people and administrative staff are brought together to work as a dynamic 'tiger team' for developing new products.

The results have been phenomenal. New products have been introduced in a fraction of the time, product quality becomes radically better and the design is 'right the very first time' because the range of skill contributing to the design ensures that all aspects are fully considered, and the company introduces products, customers want to buy.

To explain briefly, the objective of JIT manufacturing is to change the production process so as to eliminate waste. A lot of emphasis is placed on eliminating inventory, not only because inventory is an expensive waste in itself, but also because high levels of inventory have traditionally been used to hide other problems. Vendor delivery problems is an example of this syndrome. It has been an accepted practice for vendors to occasionally deliver materials later than scheduled. A traditional manufacturer will hold additional safety stock of those materials so that a late delivery will not disrupt production. This inventory method promotes waste and is a prime target for a JIT approach.

Although an ill-defined term, JIT usually incorporates at least the following approaches:

- Change in the shop-floor layout to reduce movement of materials.
- Reduction in production setup times so that products can be made in very small batches (ideally a lot size of one)
- Synchronization of the manufacturing process so that subassemblies and components are available just when they are needed and not before (or after)
- Creation of mutually beneficial relationships with suppliers.

15.4 SHOP-FLOOR LAYOUT

A traditional manufacturing plant is usually not laid out for JIT manufacturing; rather it divides the factory into machine centres that group together machines of similar type. Jobs are processed by moving the semi-completed material from one machine centre to another according to the production routing of the product. In a typical traditional manufacturing plant, the actual run time is frequently less than 10 per cent of total production time. The rest of the time is taken up by queuing and material handling.

This approach requires that the product should be moved physically a considerable distance during production as it is transferred from one work centre to another. Movement is not only wasteful in itself, but also leads to material damage, high inventory investment and long production cycle times. Another problem is that this layout leads to complex control procedures. When a production route is complex and the job takes several days to compete, complex planning and control systems are necessary to enable managers and supervisors to keep track of production schedules.

JIT layout minimizes the movement of materials, people and tooling. This system usually involves layout out of the shop floor as a series of manufacturing cells. In a cell system, dissimilar machines are grouped together based upon the process to be performed within the cell. Instead of a batch of materials being moved between several work centres, a number of different machines are grouped into a cell so that all production tasks can be completed with virtually no movement of materials.

Such a grouping of activities assumes a degree of repetition. Techniques of **GT** are employed, where products are categorized according to production processes. Cells are dedicated to the production of a group of products and contain all machines, assembly areas and test equipment necessary to make that specific group of products.

In addition to eliminating unnecessary material movement, this layout style lends itself to flexibility. The rate of flow through a cell can be changed easily as customer orders change. People within the cell are trained to perform multiple production tasks and can be assigned to a wide range of activities based on current needs. A cell can also become a responsibility centre where production personnel, control their own quality, schedules, material planning and job assignments.

15.5 SETUP-TIME REDUCTION

Setup-Time Reduction has a major goal in its own right provides benefits in lot-size reduction, shorter lead times and improved flexibility. Reduction of batch sizes usually requires a reduction of setup times. A **JIT** manufacturer spends a great deal of time and effort in studying setup times and devising methods of reducing them.

Traditionally, large production batches cause high levels of inventory—both in **WIP** and finished goods. Batch sizes (or lot sizes) frequently are determined by setup times. If eight hours are needed to set up a machine, then making a large quantity after doing all that setup work is generally desirable.

In contrast, a prime objective of JIT manufacturing is to reduce inventory not only because inventory is costly and wasteful in itself, but also because high inventory levels traditionally have been used to cover a multitude of other shortcomings. High levels of buffer stocks are used to alleviate problems resulting from such situations as bad scheduling, poor vendor delivery, poor quality and inaccurate record keeping. An important tool for the reduction of setup time.

Setup time reduction cannot be achieved by one or two changes. Reductions are achieved by a concerted systematic effort throughout the entire production process. The main offenders are identified first-those processes with longest setup times and largest batch sizes. Experience proves that setup times can be reduced by approximately 50 per cent by merely stalling the processes and by establishing better procedures such as: to setup with people who do the tasks, videotaping the process in action and by analyzing the activities step-by-step. Once initial studies have been completed and changes implemented, setup time can be further reduced by modifying dies, tools and jigs. Techniques and innovative ideas associated with these programmes have names like **SMED and OTC**.

In some cases, achieving the level of setup time reduction required for **JIT** manufacturing while using existing equipment is impossible. New production equipment must be purchased specifically for the purpose of reducing setup times throughout the entire production process. Typical of obsolete or non-adaptive equipment are heat-treatment ovens, wave soldering baths, extrusion or pain processing lengthy wash-downs and some specialized test equipment.

Setup time reduction can have a significant effect upon **WIP** inventories, cycle times and production flexibility. Benefits are achieved through many small improvements over an extended period and for this to be meaningful, analysis and monitoring methods are needed.

15.6 AGILITY OF AGILE MANUFACTURING: SOME EXAMPLES

Agility requires developing capability within an organization to naturally and quickly adapt overtime to changing market needs and customer requirements. This adaptability must be accomplished without compromising quality or increasing the costs. Change must become the norm, an opportunity to thrive, but not a threat. Agility is about change. Like agility itself, change is certainly not new. The only companies that have not changed are the ones that are no longer with us !

A firm cannot be fat and agile at the same time. Before it can get agile, it has to get lean in all aspects such as lean product development, lean manufacturing, lean vendor arrangements, lean staff, etc. Any organization that has not mastered 'lean' manufacturing should not expect to have too much success in pursuing 'agility'.

There is certainly some benefit in dropping out of the race towards increasing sophistication and automation. If we are able to do a good job with simple, reliable, cost-effective processes and equipment, we have achieved our goal and certainly not when we introduce sophisticated technology-driven solutions, even if they are world's firsts. There should be means of building flexibility into our manufacturing equipment to make them easily convertible at low cost when product structure changes.

Agile manufacturing is a system characterized by its ability to allow rapid response to continuously changing customer requirements. Customer requirements define the **FRs** (Functional Requirements) of a product design. Any change in product design requires corresponding alterations in process design. The modified process design can be executed quickly only if the configuration of the automation system design can be changed in a short time. An automation system comprises machine tools, and handling devices, typically robots with dexterous hands. Therefore, to respond to a quick change in customer requirements, the machine tools and handling equipment will have to be flexible and/or reconfigurable. By flexibility, we mean complete re-programmability without any change in hardware. By re-configurability, we mean provision for change in configuration by quick repositioning of the hardware components, or replacement of end effectors. This discussion deals with re-configurability of automation systems, as it has a significant role in the promotion and success of Agile Manufacturing. Three main areas related to re-configurability of automation systems, where research is being conducted at **RAL** are: development of reconfigurable robotic arms, **DFAM** and development of a dexterous manufacturing hand.

This requires development of modular joints which can be quickly reassembled with the links to provide numerous configurations, systematic methodology for design of an automation system and its evaluation; the methodology facilities forward and backward, flow of design data between the three phases of design, namely, product design, process design and automation system design. Conceptual design of a novel dexterous manufacturing hand which integrates manufacturing and handling functions.

Product inspection has played a vital role in fulfilling the productivity goal of many manufacturing organizations in the face of increased automation. Continuous demand for minimizing the number of defects in job production along with the advancement in the state-of-the-art microcomputers and vision system would provide a non-contact, non-destructive method of inspection at a faster rate, thus enabling 100% inspection to be effectively implemented in industries. An image of the manufactured product/part as grabbed by the vision camera has to be electronically obtained and stored in the host computer in digital form. The computer, with the help of various developed software would analyze the digitized image and carry out the necessary mathematical as well as statistical calculations so as to describe various product features as connotated by the shape number of the objects. The description of the objects through the shape numbers has been found to be independent of the size, orientation and position of the objects. The order of the shape number would indicate the precision with which that number would describe the pattern of the given object. The concept of the degree of similarity between two objects has been introduced so as to find out the closeness of the two objects with respect to their shapes. Dissimilar objects have a low degree of similarity, whereas analogous objects would have a high degree of similarity. Attempts have been made to extract the shape numbers of given objects and to evaluate the degree of similarity between unknown objects from the point of view of their shape numbers.

Right from the emergence of the competitive era, **TQM** gained a momentum in academic and manufacturing arenas. In continuation of this, many techniques such as **QCs**, **FMEA**, **QFD**, became popular. Among these techniques, **QFD** has received significant attention among both researchers and practitioners. While **QFD** is found to be a powerful mechanism for transferring customers' pulses into design and production functions, its scope does not extend to all other functions such as marketing, maintenance, sales, etc. Considering the inadequate scope of **QFD**, we may design a refined **QFD** model under the name **TQFD**. In essence, **TQFD** covers **QC** activities that are undertaken at different functions of the enterprise.

Even though **TQFD** ensures the transfer of customers feelings and view in totality, its successful application demands high level of expertise. However, obtaining expertise by appointing experts and consultants always proves to be more and more expensive and time-consuming task. Hence, the elements of information technology are required to be exploited.

To study the global environment and thereby inspecting the stipulated feature, Nitzan, David (1981) and Baird, Michael (1978) have suggested several instances where vision systems are applied for searching the particular image segment in a given interval inside a vision envelope. Dyer Charles (1983) used the Hough transform technique for finding out the position orientation of the image segment in a vision envelope.

15.7 FORMATION OF MODEL

The process of formation of a model and analyses of an image is incorporated in the vision strategy to be employed. It is essential to frame a model of the scene to be analyzed for searching an image segment. This is done every time a new scene is searched. This procedure involves fixation of the tolerance zones for the scene, which is to be searched for the likely location of the image segment. Details like height, width and breadth of the scene are required for locating the tolerance bank in the scene. This band is created with the 'ON' pixels in the binary image processing.

Agile manufacturing incorporates a sophisticated AVIS in order to reach the target of inspection.

Decision-making in the area of selection of suitable material handling equipment has been influenced by the ongoing development of new technology, practices and equipment having enhanced capabilities and complex characteristics. It can be emphasized that the specified problem of material handling equipment selection under specified critical and operating conditions has been a complicated task due to the presence of many feasible alternatives and the need to fulfil various conflicting objectives. In selecting the handling equipment, the experts in this area should constantly keep in mind the fact that successful solution to most material handling equipment selection problems would involve, within the constraints of a proper and suitable matching of optimal solution of existing or contemplated physical facilities and environment and also against related cost trade-offs. Hence, under these constrained and conflicting circumstances, the application of **KBES** has been found to be the plausible strategy for the optimal selection of material handling equipment and thus widening up of the potential effectiveness of this technique in mechanizing complex knowledge intensive decision-making problems. The present discussion focuses the module of development of an expert system for selection of material handling equipments under given critical and operating conditions with the present-day material handling needs in industries and other areas of requirements.

Introduction of electronic computer has caused drastic changes in almost all the fields of men's activities. In manufacturing, it has led at the beginning in developing and extensive usage of computer controlled machines and manufacturing systems. The use of **NC** and **CNC** machines has caused important improvements in quality and quantity of the products but has caused also a lot of new organizational problems. The real challenge came in 1970. The oil crisis, declining demand and oversupply, together with emergence of strong **Japanese** competitors, forced traditional manufacturers to search for new organizations of manufacturing. Short series production succeeded over large series manufacturing. Better quality cost-effective products and quick response became the market needs. The dissemination of computers not only in manufacturing departments like **NC, CNC, CAD, CAM** or **CAQ** system, but also in forecasting, planning, re-planning and controlling areas of the companies, together with developments done in simulation and data analysis methods has brought foundations for the real flexible manufacturing–agile manufacturing.

15.8 MEANING OF SIMULATION

Simulation is an expression which is well known and used by many people, at least in a general way. Dictionary definitions suggest that simulation involves imitation or mimicking, giving the appearance or effect of some elements. As a technical term, simulation will be defined as imitation of the behaviour or a real-world process or system over time. The system may be natural or artificial, in existence presently or planned for the future. For simulating purposes, a model of the previously defined real-world system has to be developed and its behaviour under various conditions analyzed. Simulation is not a new method, but the wide availability of the computers has opened new wide horizons for this technique.

One of the missing links between **CAD and CAM** is the virtual absence of any systematic methodology for generating and evaluating the alternative ways to manufacture a proposed design.

Most integrated **CAD/CAM** systems try to generate a single process plan for a given design, but in general there may be several alternative ways to manufacture the design. How easy it is to manufacture or whether it is even possible to manufacture the proposed design at all, may depend on which alternative is chosen. Thus, these alternatives should be generated and examined, to determine how well each one balances the need for a quality product against the need for efficient manufacturing.

Generating consistent and flexible processes to suit a dynamic production environment is an important objective. Generative **CAPP** systems based on features are being explored by researchers and practitioners. Features represent a collection of geometrical entities in an intelligent form and hence provide information at a higher conceptual level than is available in traditional **CIM** which holds the key for agile manufacturing applications. Advances in information technology have resulted in tremendous reduction in the cost of implementing **CIM** applications are likely to grow exponentially. Indian Ordnance Factories have made continuous efforts to integrate advances in IT manufacturing process. Experience gained in an environment for introducing various agile manufacturing applications in machining industry (using FMS and computers in integrated form for managing business operations) have resulted into formulating an optimum strategy for integration. **CIM** handles all aspects of business right from marketing research, order processing, materials management, manufacturing, sales and distribution to handling post-despatch service complaints. At present, environment in customerizing the product has become a compulsion. **CIM** offers a distinct advantage in such environment.

Concerned research and development efforts initiated by enterprises interested in the acceleration of product development have led to innovative technologies like **CAD, NC, RPT** and **CE**. These technologies are aimed at improving and hastening the manufacturing of protypes. Contrary to the other conventional fabrication techniques which remove unwanted material to get the final product, **RPT** is an additive process which builds the three-dimensional physical objects by joint particles or louvres or materials. **RPT** is also called 'tool-less manufacturing' to emphasize its direct transformation of an abstract **CAD** file into physical objects.

Though several variants of RP technology like Fused Deposition Modelling, Selective Laser Sintering, Laminated Object Manufacturing, Solid Ground Curing are available, **SL** is the most widely used method due to several functional advantages like greater accuracy, better surface finish and transparency. The experience of gas turbine manufacturers has proved that photoelastic model preparation time can be shrunk significantly by substitution of conventional manufacturing process with **SL**.

Here it is presented the **COVIRDS**, a tool for creating, edition and visualizing three-dimensional (3D) shapes in an immersive **VR** (Virtual Reality) environment. **COVIRDS** provides initiative and efficient voice and hand-based inputs for interacting with product designs. While existing product designs can be directly visualized in **COVIRDS**, new product shapes are easily created via voice (and/or 3-D menu-based commands and hand action/motion in the three-dimensional (3-D) space. The research consists essentially of two components:

(i) the first centring on the human-computer interaction (Interface Definition) and
(ii) the second on geometrical representation for visualization and editing (Shape Representation).

The interface definition research focuses on identifying the most appropriate interaction modalities (voice/one-hand/two-hand) for various tasks in a **VR-CAD** environment. Shape representation research focuses on developing representations suitable for interactive editing of parametric and free-form shape geometry.

In the late 1960s, **AI** gained popularity. **ES**: one of the major areas of **AI** has achieved considerable success in recent years and these are widely used in many areas of engineering. In production engineering it can be used in various fields like tool and material selection, creation of machining data bank, estimation and costing, maintenance, sales and purchase scheduling, etc.

As the cost of computers is reducing with an increase in their speed, power and sophistication, we can find the utilization of computers in many manufacturing activities. This is true after the rapid rise in the usage of **CNC** machines for manufacturing. The present trend in production engineering is to automate the activities and reduce human intervention. During **CNC** programme generation, optimum machining parameters should be given a great need to meet production engineering

The data bank created is a stand alone system. Generally, the data bank can be divided into types:

(a) Storing all the data in memory and retrieve systems
(b) Systems using empirical equations
(c) Systems using mathematical models

First and second types are easy to consult and develop, whereas the third type is very complex. The data bank generated is the combination of first and second type systems.

ES are software programmed or it is a computer based system that can perform some task which requires expertise. This is also referred to as **KBES**. **ES** tools are nothing but software systems. Number of tools are available for building expert systems. These can be broadly classified into two types.

- The first type is designed specifically for programmers and it requires thorough knowledge of **ES** development languages **(Lisp and Prolog)**.
- The second type is aimed at individuals and already developed softwares like **PCPLUS** are used for individual application. To create the machining data bank, the second type **ES** development tool **PCPLUS** is used. It is a highly functional tool for developing **ES** on **PCs**. It offers a sophisticated frame structure.

As a result of continual optimization of costs, quality and time, new trends may shortly replace the old structure in the area of product development processes. In other words, the integration of external partners and suppliers in several locations in the development process as well as enhanced computer capabilities under high speed networks is dramatically changing the world of engineering. **CE** is one of the key factors for the emerging next step of industrial revolution initiated by **IT** and globalization.

Under these circumstances, several enterprises have proposed some new solutions using **CE** technology. For example, some companies have developed network systems covering several production sites worldwide. Those systems are based on **DMU** technology (based on the concept of **CE**) and data sharing and parallel processing are available in the global network. In Japan, **EXEN** achieved the **FCR** project, which has developed **3D CAD/CAM** system to integrate all of the experimental data to implement **CE** technology. But application of **CE** technology in these systems so far is mostly limited to development of network systems, quality control of products, parallel processing of each division of the production department. In fact, many departments are involved in network systems.

In the mid-90s, it was first realized (formally at least) that 80% of the final cost of product was frozen by decisions made in the early design stages. More importantly, it was discovered that it costs two orders of magnitude less to make changes in the part design, during the early stages than in the later stages of prototyping/manufacturing. The main causes for these are:

- The sequential nature of the product development process
- Lack of automation in transition between tasks in product development

As concerns of lower cost and faster lead times became overriding, the above observations lead to the concept of **CE**. In simple terms, **CE** implies simultaneous incorporation of the concerns of downstream modules/tasks such as manufacture, assembly, maintenance, in the design stage itself. It refers to the concurrent design of products and the processes required to realize the same. That it was hoped would reduce the iterations through each of these tasks in the present sequential process of product development.

Continuous improvements in existing design in manufacturing processes have been the hallmark of all leading companies in the manufacturing industry to help them deal with the ever-increasing competition that has been manifesting itself in the last few years. The rising expectations of customers from the markets and the saturation of the market with basic products has led to greater competition, with the result that customers have started demanding greater product variety and higher quality (including long-term reliability) at lower prices and shorter delivery lead times.

How could these challenges be tackled effectively? Obviously, the requirements of more variety and better quality has put pressure on the manufacturing costs and delivery times, creating a seeming impasse. This is when industrial engineers have started looking at the processes of design and manufacture more closely for a solution and this has resulted in bringing to light some of the basic issues, which have to be addressed to resolve the situation, namely:

- Sound design and manufacturing engineering, which require well-informed staff and systems which maximize their effectiveness among other things, by enabling them to draw on previous experience and exchange information with others
- Better interaction between the design and manufacturing engineering departments to ensure the manufacturability at the design stage itself, because any corrections to be done later would become progressively more expensive
- Less number of parts for better quality, which would mean great complexity of the parts requiring more sophisticated machine tools and designers and manufacturing engineers with better knowledge
- Better control of the manufacturing process through effective process plans which can be generated in a short time by drawing on past experience.

Although non-verbal communication is getting more attention lately, it is difficult for a distance learning system to take this into consideration, because in distance learning, the points of focus should be more exactly and fully shared between a teacher and students. There are many TV conferencing systems available today, but they are developed for conferencing and not for education. Although conferencing is also an interactive process among participants, the primary goal is to find solutions collaboratively for upcoming problems.

Thus, these processes are to be constructed. Education, on the other hand, usually has a more specific goal to provide students with a clear perspective and understanding of well-structured knowledge and experience. Thus, in distance learning, mechanism that permits a teacher and students to share the perspective and point of focus is important. Non-verbal communication is important if we consider education in Asian countries, where large portions of knowledge and experience in manufacturing is associated with motional skill. If we consider an assembly, for example such motion should be taken into account if we wish to produce an adequate design plan. So the subject of transferring motional skill is important not only from the standpoint of training, but also from the standpoint of establishing collaboration between different engineers.

Therefore, we should develop a prototype system for supporting skill training with special attention paid to the interaction between a teacher and student. The demonstration deals with the skill training of penmanship. The system measures the position of the pencil of students in 3-D space and compares those with those of the teacher and gives audible instructions.

Reverse engineering by scanning is a process of quick reproduction of parts without resorting to preparation of drawings for creating a replica or a geometric variant of the same and quick preparation of moulds directly from the finished component. Several devices can be employed for capturing the data of a model and data capture can be done irrespective of the material of the model. Depending on the features of the scanning device, scanning can be done more closely where complex shapes are present. The data so captured are used to create a surface/solid model for further analysis and modification and subsequent tool path generation for manufacturing the component. Usually digitized data will be in the form of huge set of points from which the user will be able to create a model automatically and generate a **CNC** programme for respective machines for machining the component. The output from scanning could also be converted to **STL** files and input to rapid prototyping for quick realization of the product concepts. Modelling and tool path generation from scanned data is done with the help of a software usually interfaced with the data capture device. The captured data could also be transported to a **CAD/CAM** system for further processing. Reverse engineering is a powerful **CAD/CAM** tool for component manufacture with a very short lead time.

In today's complex manufacturing environments, robots are playing a greater role, than ever. They assist in the production process on the assembly level, but also at more flexible levels such as parts supply and distribution, task distribution and beyond. No longer do we deal exclusively with stationary robots; mobile robotics is becoming more common. The environments that such mobile robotic systems must navigate are very complex in nature and require special arrangements be made to allow them to navigate or manoeuvre, successfully and safely. The stationary, multi-axis robots commonly used for assembly tasks require little use of sensors as their environment is controlled and their task movements are pre-programmed into the system. Mobile robots experience the industrial surroundings they move in and by this very nature must be able to react to changes in their environment, both expected and unexpected, and able to interact in an intelligent way with their environment. To accomplish this task, mobile robots must carry systems of both sensors and actuators on board, which are designated to sample their surroundings and allow the robots to influence their surroundings.

Manufacturing industries now can access information across their organizational boundaries. Some of the information they need for their manufacturing, marketing of their products, and management of their finance and resources are maintained reliably by the other involved parties. Hence, a suitable architecture is now required to access the information via reliable networking methods, and integrate the acquired information into application systems that help the production and business processes. Much work has been carried out on how we can integrate multiple application systems, by linking related data items from various applications. We can access equivalent and related information from multiple sources of suppliers, distributors, etc.

Today, the nature of technology, people, and market are changing rapidly. The methods and practices of the past and, in particular, mass-production of the iron or craft age, are no longer valid. A recent trend is towards an **IT**-based 'invisible enterprise' where key functions and services are all subcontracted to a set of selected suppliers. The time is not far when an enterprise will be totally virtual (from **IT** integration sense–merely serving as a glue, typing all the required functions or processes back to an organization's value system. There are many enterprises today in which at least one of the following life cycle functions is outsourced (meaning performed outside over a global IT-based collaborative network: Analysis; Design; Hardware-Prototyping; Manufacturing; Marketing or Customer Support). For example, an enterprise may choose to design its product in-house, manufacture it somewhere else, and rely upon a third party (for instance, a **VAR** for marketing and servicing their products. The single rifle-shot approach of the past–that in trying to get everything right through marketing–has not worked well. People gain a collective perception of a product quality by its actual field use, by its value to their lives, and/or by relative comparison to other smaller products. Although, motor vehicles may be produced in one region and sold in another, the regional business unit is ultimately responsible.

The increasing integration of Indian economy with global markets has registered a quantum jump; however, the situation from a technological point of view remains far from satisfactory. Common use of the state of art technology and upgradation of 'in-use' technology have not been taking place as a continuing or regular activity. With large industries increasing their outsourcing requirements, it is the **SMIs** which now have to access advanced manufacturing technology for continuous supply of high quality components and materials to large units.

SMIs usually do not have appropriate information about and access to the latest technologies due to financial, technological and management constraints. Only some cases, say, ancillary **SMIs** receive technological input from their mother units. Hence, the national laboratories, research organizations, institutes and technical universities have a crucial role to play in helping these **SMIs** in adapting the state-of-the-art technologies. Discussion on this subject focuses on the current scenario on technology transfer from technical universities to industries. It critically reviews existing models for interfacing university-industry interaction across various countries.

Industrial training is the panacea to the causes of human error. When designing a training system for providing a commendable productivity enhancement, industrial training is one of the key aspects of industrial activities. The training analyst is faced with challenges in the design of training systems/programmes. It is not an infrequent occurrence for the training analyst to be required to define operating procedures for a new system in order to allow the definition of the training requirements to be achieved. The challenge for a training analyst is thus to provide a definition of what tasks require training, and what instructional strategies should be employed to provide consistent levels of skilled performance. The necessity of continuous training been frequently emphasized in guides for the implementation of **TPM**.

15.9 PROBLEMS OF INTRODUCING TRAINING PROGRAMMES

Careful analysis of the interview with more than 100 captains of industries reveal that the main problems associated with introduction of training programmes are:

In recent years **GT** has become a popular approach for improving the productivity of manufacturing systems. It is based on the principle that 'similar things should be processed similarly'. In the context of manufacturing these, similar things would include design, production planning, fabrication, assembly, etc. Group machining methods constitutes a novel scientific approach to the planning of component machining techniques. Theoretical research in GT lags behind its practical applications. Many industries have been attempting to fully realize the benefits of **GT**.

In **GT**, similar components are identified that require processing on a set of machines and in turn these machines are grouped together to form a 'Cell'. A cell can be defined as a collection of dissimilar/similar machines dedicated to the manufacture of similar type of components. The cell system has been recognized as a more efficient production system for batch size production. Design of machine cells is a very complex phenomenon. Though many methods are available in literature published earlier, a straightforward and efficient method for formation of machine cell is yet to emerge.

It is hard to predict the future in advance particularly with engineering-driven industries like manufacturing. In world class manufacturing enterprises, manufacturing will not be like what is today, but it will be **GREEN** due to the demands of society.

According to **David Gardiner, Assistant Administrator, Policy Planning and Evaluation, US Environmental Protection Agency**, environment-friendly products have become not only a powerful marketing tool across all sections, industries and services, but also a tool that recognizes customers' preferences for products that have less harmful impacts on the environment.

Minnesota State, USA estimates six manufacturers using recyclable materials created around 1700 jobs, 39 million dollars in new wages and an increase of 100 million dollars in gross state product.

Although the concept of different types of product realization have been discussed, how do we define product realization? Product realization is the process by which we develop a product from the conceptual stage to the point when the product is released to the market with a minimum amount of time and effort.

Taxonomy (classification) is the additional dimension added to product realization so that the product realization can be applied in an implementation independent manner. Thus, the concept of product development can be applied to any company and the steps for implementing it will remain same. The scope and content of taxonomy for product realization are implicit in the following.

- Types of information and sources
- Complexity of tasks
- Degree of incompleteness and ambiguity
- Timing and decision-making
- Order of decision-making
- Communication mechanism

Throughout history, manufacturing is going through successive periods of great changes. New materials, new techniques, new technologies have always been at the root of these changes. Currently globalizing markets, demanding customers, congenial environment for foreign investment and other factors such as these have made "manufacturing" more complex. In the current environment, manufacturing has to be interpreted as consisting of a chain of all activities, namely, conceive, develop, manufacture, distribute, sell, maintain and dispose the products while following regulations on environment and keeping internal and external customer satisfied.

These are called CE Taxonomy in Six Dimensions as elicited by Prasad.

- permits many parts of the product to be designed in parallel
- helps in breaking down the concept into smaller, projects and integrating the projects into realized product
- helps in developing products and services required by the customer
- enables CE beforehand to know potentially complex tasks which are difficult to coverage
- provides the means for the management to "hammer it out" with core teams

Every organization gives priority to the fundamental parameter, which is the demand, when choosing its strategy even before the surveillance of competitors and changes the productive manufacturing systems. Continuous analysis therefore provides a vital insight into changes that are required in the productive manufacturing system, namely, design and production. The problem during the transformation process culminates in a number of design changes. During the transition from design to execution, the time of delivery of a product, i.e., the **MLT** governs the capability of a firm to quickly satisfy the demand. The reworking effort required and scrap amount are major problems which augment the **WIP** inventory because of an escalated manufacturing lead time.

The various production philosophies need to be discussed starting with the craft production in which products were designed and built on order by the customer.

Mass production followed the exploitation of the economy of scale and large quantities to satisfy broad customer needs at lower prices.

Lean production aims at exploiting the good prints in production concepts, i.e., a high-quality custom-made product at price that could be possible by mass production. Since the basic concept of lean production is to minimize waste, it encompasses all phases of product development and thus can be considered a production philosophy by itself.

An agile company is basically a lean company with the additional quality of adaptability to any future changes in the industry.

A virtual corporation is "a network of independent firms that share costs, skills and market across through **IT**".

16
Benefits of Agile Manufacturing

The message of this book is novel, simple and exciting. Novel, because it heralds a new direction for technological progress. Simple because it offers a straightforward quantitative formula for it. Exciting because it is profitable. The message is also radical, it could change the future.

Factor Four describes a new form of progress, resource productivity, one which meets the overriding imperative for the future, namely, sustainability. It shows how at least four times as much wealth can be extracted from the resources we use. As the authors put it, the book is about "doing more with less; but this" is not the same as doing less, doing worse or doing without.

In 1972, the Club of Rome published 'Limits to Growth', which sent shockwaves around the world by arguing that we were rapidly running out of essential resources. This new Report of the Club of Rome offers a solution. It lies in using resources more efficiently, in ways which can already be achieved, often not at a cost, but at a profit.

This book contains a wealth of examples of revolutionizing productivity, in the use of energy, from hyper cars to low-energy beep; materials, from sub-surface drip irrigation to electronic books and transport, from videoconferencing to Cyber Tran demonstrating how much more could be generated from much less, today. It explains how markets can be organized and taxes rebated to eliminate perverse incentives and reward efficiency, so wealth can grow while resource used does not.

The benefits are enormous: profits will increase, pollution and waste will decrease and the quality of life will improve. Moreover, the benefits will be shared: 'progress' will no longer depend on making ever fewer people more productive. Instead, more people and fewer resources can be employed. While for many developing countries the efficiency revolution may offer the realistic chance of prosperity within a reasonable time span. The practical promise held out is huge, but it is up to each of us, as well as to business and government, to take it.

Doing more with less is not the same as doing less, doing worse or doing without. Efficiency does not mean curtailment, discomfort or privation. When several presidents of the **US** proclaimed that 'energy conservation means being hotter in the summer and colder in winter', they were not talking about energy efficiency, which should make us more comfortable by improving buildings so that they provide better comfort whilst using less energy and less money. To avoid this common confusion, it is better to avoid the ambiguous.

17

How Vital is Resources Efficiency?

Although they are somewhat abstract, there are seven compelling reasons for achieving resource efficiency:

(i) **Live better:** Resource efficiency improves the quality of life. We can see better with efficient lighting systems, keep food fresher in efficient refrigerators, produce better goods in efficient factories, travel more safely and comfortably in efficient vehicles, feel better in efficient buildings and be better nourished by efficiently grown crops.

(ii) **Pollute and deplete less:** Everything must go somewhere. Wasted resources pollute the air, water or land. Efficiently combats waste and thus reduce pollution, which is simply a resource of a place. Resource efficiency can greatly contribute to solving such huge problems an acid rain and climatic change, deforestation, loss of soil fertility and congested streets. By themselves, energy efficiency plus productive sustainable farming and forestry practices could make up to 90 per cent of today's environmental problems virtually disappear, not at a cost but given favourable circumstance at a profit. Efficiency can buy much time in which we can learn to deal thoughtfully, sensibly and sequentially with the world's problems.

(iii) **Make money:** Resource efficiency is usually profitable. You do not have to pay now for the resources that are being turned into pollutions and you do not have to pay later to clean them up.

(iv) **Harness markets and enlist business:** Since resource efficiency has the potential of being profitable, much of it can be implemented largely in the marketplace, driven by individual choice and business competition, rather than requiring governments to tell everyone how to live. Market forces can theoretically drive resource efficiency. However, we are still confronted with the considerable task of clearing barriers and reversing draft incentive structures that keep the market from working fully.

(v) **Multiply use of scarce capital:** The money freed up by preventing waste can be used to solve other problems. Developing countries in particular, with loss of their capital sunk in inefficient infrastructure, are in an excellent position to multiply the use of scarce capital. If a country buys equipment to make very energy efficient lamps or windows, it can provide energy services with less than a tenth of the investment that would be required to buy more power stations instead. By also recovering that investment at least three times faster and reinvesting it elsewhere, the services rendered by the invested capital can rise more than 30-fold. Calculations show savings can even be far higher. For a good many developing countries, this could be the only realistic way to achieve prosperity in a reasonable time span.

(vi) **Increase security:** Competition for resources causes or worsens international conflict. Efficiency stretches resources to meet more needs, and reduces unhealthy resource dependencies that fuel political instability. Efficiency can reduce international sources of conflict over oil, cobalt, forests, water—whatever someone has that someone else wants (some countries play in military costs, as well as directly, for their resource dependence mission is getting or keeping access to foreign resources). Energy efficiency can even indirectly help block the spread of nuclear bombs by providing cheaper and inherently non-military substitutes for nuclear power plants and their dual-purpose materials, skills and technologies.

(vii) **Be equitable and have more employment:** Wasting resources is the other face of distorted economy that increasingly splits society into those who have work and those who don't. Either way, human energy and talent are being tragically misspent.

Yet a major cause of this waste of people is the wrong and profligate (wasteful) thrust of technological progress. We are making even fewer people more 'productive', using up more resources and efficiently marginalizing one-third of the world's workforce. We need a rational economic incentive that allows us to employ more people and fewer resources, solving two critical problems at the same time. Businesses should sack the unproductive kilowatt-hours, tonnes and litres rather than their workforce. This would happen much faster if we taxed labour less and use resources correspondingly more.

Efficiency is a concept as old as the human species. Much human progress, in all societies, has been defined by new ways to do more with less. A great deal of engineering and business is about using all kinds of resources more productively, but for the past 150 years, much of the technological effort of the industrial revolution has been devoted to increasing labour productivity, even if that required more generous use of natural resources. Recently, however, resource efficiency has undergone such a conceptual and practical revolution that most people have not yet heard about is its new potential.

Progress in making resource efficiency bigger and cheaper is thus quite dramatic. It is, in fact, somewhat comparable to the computer and customer electronics revolution, where everything is continually becoming smaller, faster, better and cheaper.

Yet today, not only are there new technologies, but there are also new ways of linking them together so that in principle big savings can often be had even more cheaply than small savings. When a series of linked efficiency technologies are implemented in concert with each other, in the right sequence and manner and proportions (just like the stages in a good recipe), there is new economic benefit to be reaped from the whole that did not exist with the separate technological parts.

This is a startling contradiction to the everyday wisdom that 'you get what you pay for'–the better, the dearer. How can a slightly more efficient car cost more to build than normal, while super efficient car costs less to build than normal? Still, there are many reasons for this.

Conventional ways of doing things hold practice in a vice line grip. Most architects and engineers, too, are paid according to what they spend, not what they save; so efficiency can directly reduce their profits by making them work harder for a small fee (because their fee is often based, even if indirectly, on a fixed percentage of the project cost, so if the cost goes down, so does the fee).

Even with proper incentives, it is not easy to apply these new ideas about saving resources.

Advanced resource productivity requires integration, not reductionalism–thinking about the design challenges as a whole, not as a lot of disjointed little pieces. It therefore fights this century's trend towards narrow specialization and disintegration. It demands optimization, not rules of thumb. It requires a new approach to design education and practice. The systems that waste resources today are difficult to design because they are complex, but they are extremely efficient systems and are difficult to design because they are sophisticatedly simple.

The essentially cultural barriers to modern resource efficiency are just the tip of a very large iceberg of underlying problems. In attempting to save resources, we are faced with a daunting array of practical obstacles that actively prevent people and businesses from choosing the best buys first.

- The conventional education of nearly everybody dealing with natural resources, and the often insurmountable costs of replacing conventional personnel with the rare individuals who know better. This 'human factor' may actually be the biggest obstacle and the biggest part of what economists usually call 'transaction costs', the costs of overcoming inertia by taking action to change the way things are;
- Other transaction costs relate to the massive interests some capital owners have in preserving existing structures and more inertial from customers who may be simply ignorant about what levels of resource efficiency they could demand;
- Discriminatory financial criteria that often take efficiency jump over a tenfold higher hurdle than resource supply (for example, the very common insistence that an energy saving measure should repay its investment in a year or two, while power plants are given 10 to 20 years to pay back);
- Split incentives between one person who might buy efficiency and other who would then reap its benefits (such as landlords and tenants or home and equipment builders and their buyers);
- Prices that say little about, and may severely distort, actual costs to society, let alone costs to the environment and to future generations;
- The greater ease and convenience of organization and financing one huge project rather than a million little ones;
- Obsolete regulations that specifically discourage or outlaw efficiency–ranging from prohibiting taxi drivers who take a passenger into another territory from picking up anyone on the way back, to letting manufacturers' lorries carry only their own goods, to restricting buildings' window areas even when the windows are of a newer kind whose increased area always saves energy, to preferential haulage tariffs that give virgin materials a cost advantage over otherwise identical recycled materials; and
- The almost universal practice of regulating electric, gas, water and other utilities so that they are rewarded for increasing the use, and sometimes even penalized for increasing the efficiency, of resource (an unfortunate side-effect of the restructuring of the British electricity system).

17.1 CONSERVATION OF RESOURCES FOR SUSTAINABILITY

The **MRP-I** was introduced in the year 1975 and defined as a business/manufacturing practices that integrate the core activities of manufacturing like production, planning and inventory control to control production and materials flow efficiently leading to conservation of resources as in Fig. 17.1. Because of economic situation and the rapid decrease in the cost of computing for pushing "manufacturing control into limelight", a sound manufacturing management practice was introduce.

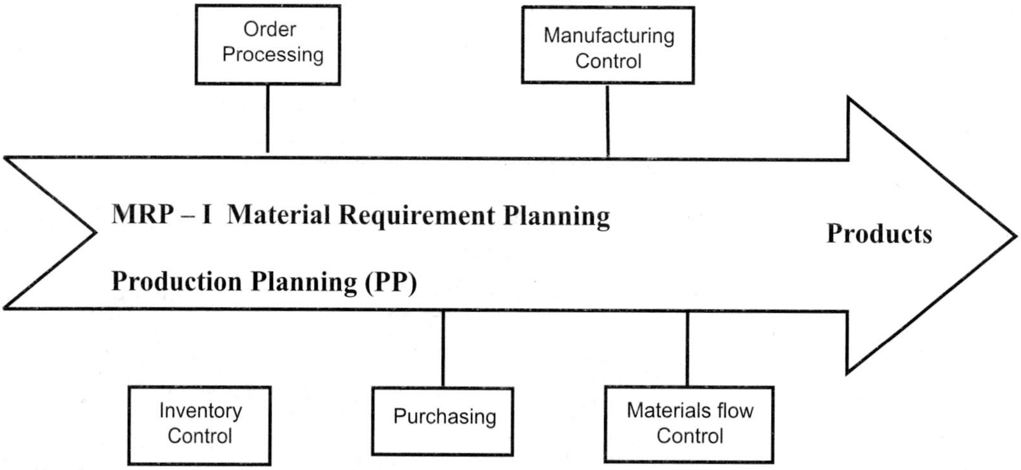

FIGURE 17.1: MRP I

The unstable interest rates, material shortages, rising finished goods inventories, order cancellations, the impact of scheduling inefficiencies on profits, etc. demanded for tighter operational control. This led to the birth of **MRP II** as a successor to **MRP I** as in Fig. 17.2. This enables Resource Planning and Customer Service leading to further Conservation of Resources. It reduced inventory by 10 to 25% and production cost with flexibility, planned operations, better control and communication.

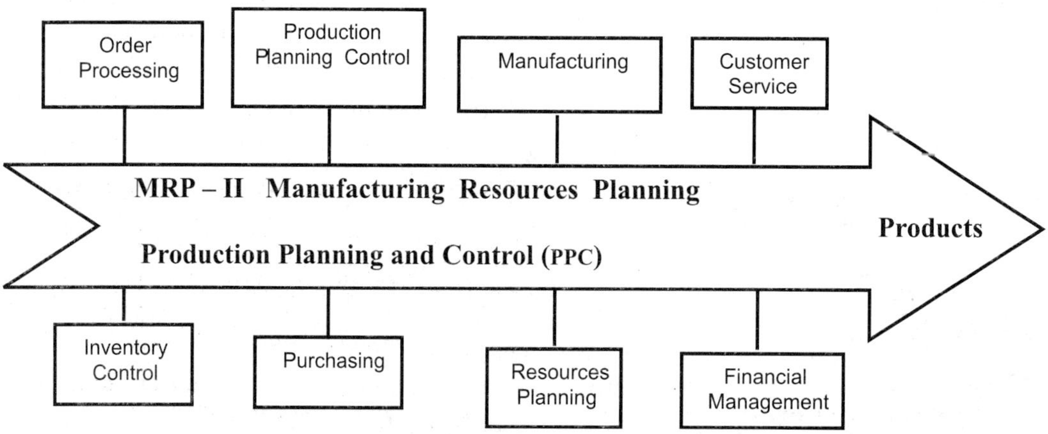

FIGURE 17.2: MRP II

After 20 years, manufacturers agreed to find an alternative to **MRP II** which has more control than that of **MRP II**. They succeeded to invent **ERP** to achieve operational efficiencies such as **Resource Planning, Lean Manufacturing, JIT, TQM,** etc. as shown in Fig. 17.3. By 1995, over 48,000 /- out of approximately 60,000 US manufacturing units implemented MRP II and tried to change over to ERP which led to higher conservation of resources.

The business/management impact of quality **MRP II/ERP** application is now extensive. **ERP** facilitates integration, error reduction, elimination of redundant operation, forecasting and planning capabilities that can provide significant cost savings and process improvement leading to :

(i) Reduced Inventories
(ii) Improved time to Market (Speed)
(iii) Increased Profitability
(iv) Significant Conservation of Resources

They are commercially available and used for the last one decade as shown in the following Fig. 17.3.

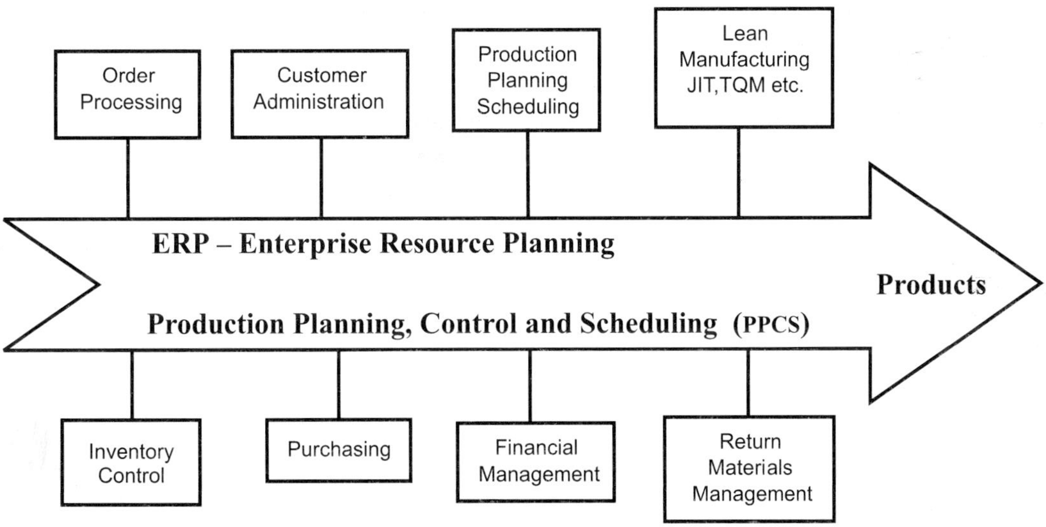

FIGURE 17.3: MRP III

As more and more manufacturers implemented **ERP**, and as customers demand increased, it is apparent that the 10-year old **ERP** will not provide a clear sustainable competitive advantage. **ERP** manufacturing practices are almost exclusively focused on Internal Execution. **ERP** optimizes order entry production planning, procurement, production, delivery and administration, i.e., all internal operations. But if competitive advantage in this century will be determined by creating and delivering customized product, the current **ERP** model is insufficient. In this decade, it is necessary to integrate customerisation with **ERP** which may be called **CERP** which is called **"Agile Manufacturing"** and is shown in Fig. 17.4 also called **CSRP**.

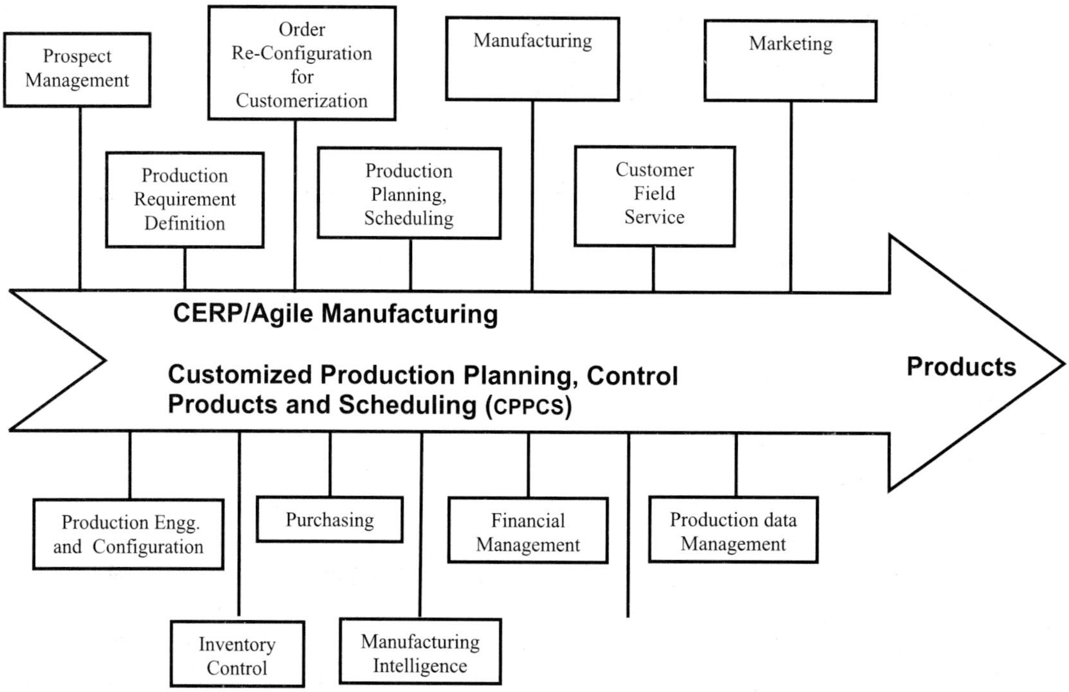

FIGURE 17.4 CERP/CSRP/agile manufacturing–synchronize the customer with internal planning and execution.

Agile manufacturing leads to a few critical business/manufacturing questions:

(i) "Which products will be customers' demand in six months?"
(ii) "Which product enhancements will create a competitive difference?

If customer preferences are changing at unprecedented rates, how is it possible to obtain critical market information? The answer is to integrate the customer into the mainstream business planning and execution systems.

CERP/AM takes the proven, integrated functionality of **ERP** and redirects enterprise planning away from manufacturing and toward the customer. It provides actionable methods and applications to create customized product by reconfigurability.

CERP/Agile Manufacturing contains four key imperatives:

(i) Optimize operations to build an efficient production infrastructure with ERP methodologies and tools; as shown in Fig. 17.4
(ii) Integrate the customer and customer-focused departments of the organization, with the core planning and production departments; and
(iii) Embrace open technologies to create a technology infrastructure that can support integration of customer, supplier and enterprise applications.

(iv) No waste of resources, as it manufactures the customized products only.

17.2 EMBRACE OPEN TECHNOLOGIES

About 15 years back, machine requirements drove software development to achieve agility. This led to Multiple Hardware Environment. The age of **UNIX** and open solutions began.

The proliferation of the PC in the manufacturing marketplace advanced both the capabilities and the expectations of enterprise applications. The need to network users and integrate enterprise applications with PC applications has driven software developers to adopt common interface standards. PC software vendors, largely driven by industry giants like **Microsoft/Wipro/Infosys** are establishing communication standards that allow interoperability of business applications. It is now feasible for applications created by different vendors with different architectures and different technologies, to integrate with each other successfully.

This ability to integrate multiple technologies and multiple applications is critical to the success of **CERP**. It is now possible to assemble disparate applications that are built by multiple vendors into one united enterprise management application. It is possible for manufacturers to give employees the technologies that can satisfy their specific business requirements and at the same time integrate with the core enterprise applications. Plant Operations, Administration, Sales, Customer Service, Field Services and Other Customer-Centric Business Functions can manage their departments with software, developed for their business work requirements and simultaneously supply and retrieve critical information from a central business system – founded on **CERP** – that can be shared by the rest of the organization.

Open systems technologies make it possible to create new strategic initiatives, like **CERP**. **CERP** asserts it in the integration of customer information into the enterprise planning and development process that will drive competitive advantage. Taking advantage of open technologies to bring customer preferences and requirements into the planning process are essential elements of the **CERP** business model which is different for different industries – big or small.

Consider the following: A sales person meets with a new customer at the customer's facility and together they discuss current and future product requirements. They discuss options, pricing and services and configure a solution to match the customer's unique requirements, a solution that no other competitor can currently offer.

Using CERP applications, the salesperson is able to record specific product requirements, commit pricing, and automatically send that information back to corporate headquarters where the product requirement information is dynamically converted into detail manufacturing and planning instructions. A production bill of materials is created, routings automatically defined, materials planned and ordered and work orders created. Thus, created critical customer information is dynamically integrated into core enterprise practices. The critical customer preference information is now stored in a centralized customer database accessible by customer service, field service, research, production planning and others. Enterprise operations are synchronized with customer requirements, i.e., agile manufacturing.

Now, consider this: The same customer uses an Internet browsers to access a manufacturer's Website to enter orders – standard or customized – any time of the day or night. The customer can modify previous orders, check on the status of outstanding orders or inquire about new options. Because this communication is integrated into the core business systems of the enterprise, planning, production and/or customer service, activities are automatically triggered by the customer's actions. Enterprise operations are synchronized with the customer.

Open technologies make both of these scenarios, and **CERP/CSRP** a reality. **CERP** requires an open technology framework which can integrate strategic departmental applications into a scalable, secure enterprise solution. Successful deployment of **CERP** is concrete and because of open technologies. A move away from proprietary systems, including proprietary **ERP** solutions, is required.

Become Indispensable in Business or Manufacturing

An organization/manufacturing industry has optimized operations, integrated the customers and implemented open Technology Architecture. The Customer-Centric Developments such as Management, Technology and Workforce are integrated into the Core Business/Manufacturing Industry now turn to ACTIONS:

(i) Create the market by your Innovations, Kaizen, quality, cost, etc., if you already own your business/industry.

(ii) If not, for new business/industry

- Identify your company's most promising, profitable markets.
- Identify markets and products that are profitable
- Predict what your most profitable markets will demand in one year, in six months, in three months.
- Generate and track leads in your most profitable markets.
- Guarantee on-time delivery to the most valued customers and all customers.
- Quote accurate delivery times on customized orders.
- Respond to customer service calls within 24 hours, eight hours, one hour.
- Profitably customize or repackage, products and services.

Implementing **CSRP**/Agile manufacturing helps you to achieve them for prosperity in all respects and **become indispensable**.

17.3 COMMIT NOW AND DELIVER ON TIME

This is possible by the application of **APS** systems. These remarkable new tools, when integrated to ERP systems, add a new ability to plan and manage manufacturing customers' requirements effectively. With **APS**, a company can immediately determine the impact of a new or modified customer order and set a realistic delivery date based on existing information and conditions. The company can ***commit now***, promising the customer a delivery date that is both real and realistic. Since the guesswork is removed from the process, manufacturers will be able to ***deliver on time*** as never before.

Traditional **MRP** is restricted by the limits of the technology that existed at its inception. Though greatly improved over the years, its structure and some basic underlying assumptions prevent MRP from advancing beyond its current capabilities. Traditional **MRP** has served the manufacturing community well for three decades and some companies may find its capabilities still adequate for their needs. But many customer-driven manufacturers can benefit greatly from the application of **APS** to their planning and scheduling needs.

APS evolved from basic **MRP**, master scheduling, resource/capacity planning and production scheduling tools through a series of technological advances and the application of sophisticated calculation methods. Advancing technology is the enabler for a radically new approach to planning and scheduling that provides a clear picture of the impact of schedule changes and management decisions in minutes making **APS** a true decision support tool for today's fast-paced and highly competitive world. **APS** is not a replacement for **ERP**. **APS** replaces the planning applications within **ERP** and leverages **ERP's** transaction handling and order management (execution) capabilities.

What is APS?

Fundamentally, **APS** represents a radical change in the way material and resource planning is executed in a manufacturing company. Traditional planning, MRP, is a step-by-step, sequential process. Ideally, material is planned without regard to capacity constraints, then the capacity plan is devised and matched to the materials plan. But the process is often not as streamlined as the designers intended. The MRP sequence encompasses multiple steps as shown in Fig. 17.5.

(i) **Enter Customer Demand** – Consolidate demand requirements, then

(ii) **Create a Master Production Schedule (MPS)** – Develop a first-cut production schedule, then

(iii) **Create a Rough Cut Capacity Plan (RCCP)** – Test the production schedule for feasibility against available plant capacity prior to gauging the materials available, then

(iv) **Validate** – Determine whether or not the RCCP plan is feasible and start over again if capacity is in question. Make adjustments, reset the production schedule and rough capacity plan, evaluate feasibility and if finally acceptable, then

(v) **Create a Material Requirements Plan** – Once a realistic production schedule and rough capacity plan are finally derived, create a detailed materials plan to determine the feasibility for all levels of the **BOM**, then

(vi) **Create a Capacity Requirements Plan (CRP)** – Once the materials plan is developed, develop the capacity plan, then

(vii) **Validate** – Check once again for feasibility and adjust as necessary if material and capacity plans are inadequate, start over again at the beginning with the Master Production Schedule; make adjustments, retest and if entire process is finally acceptable, then

(viii) **Create the Final Production Plan** – MRP, as you can see, is a top-down, single-direction, sequential process involving many potential restarts prior to resolving the final plan. During the often lengthy process, adjustments made to accommodate capacity problems may cause material problems and vice versa. It is sometimes necessary to cycle through this process several times before a complete, balanced plan results. Time and resource limitations often leave the planning process incomplete or not fully resolved.

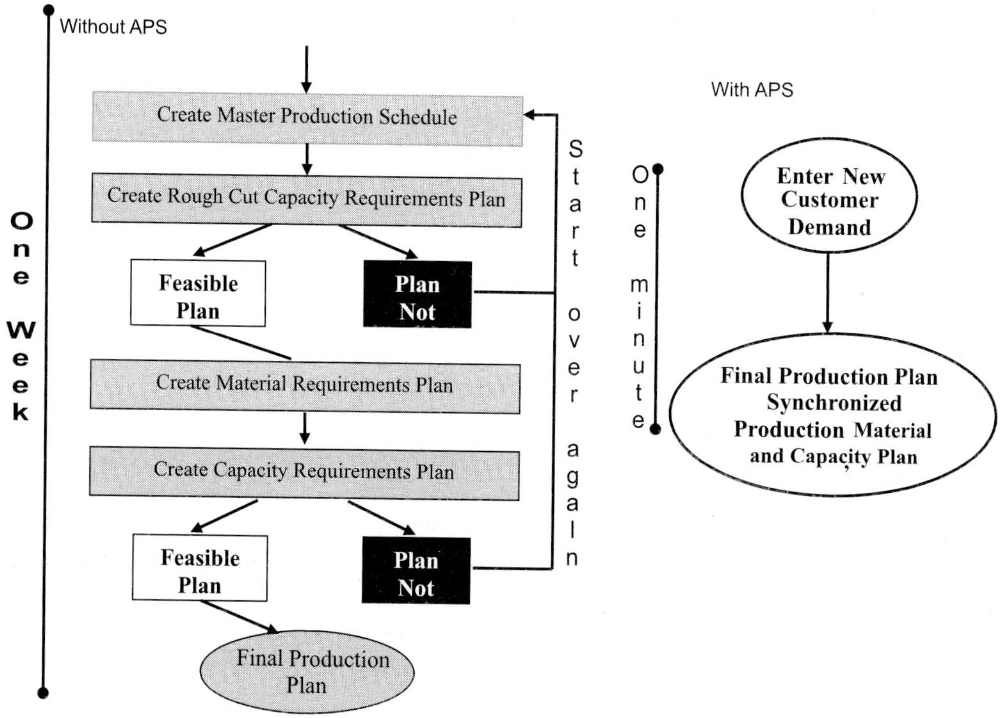

FIGURE 17.5 MRP sequence.

APS uses a finite capacity or constraint-based approach, meaning the plan will not over-commit manufacturing resources beyond available capacity. Because resources are planned at the same time as materials, there is no need to make unjustified assumptions about resource availability. Each activity is fully planned and coordinated with other demands on work centres, people, machines, etc. to generate schedules that are based on reality, not based on fixed lead time estimates and cavalier assumptions about resource availability. Flexible, not static, data is used to build the plan.

APS is fast. Exploiting recent advances in computer technology, the APS planning cycle is typically carried out immediately – and measured in minutes or even seconds – as opposed to being generated slowly and after business hours or over the weekend. The significance of this point is that planning now becomes a decision support tool, not simply a reporting tool. Resource availability questions can be answered, alternatives immediately explored, and the impact of disruptions – and proposed solutions – can be identified without delay.

18

Green Engineering

Green engineering is the study of product and process design and manufacture for environment friendliness, product functionality, quality and commercial viability. The main aim of this is to identify, develop and exploit new technologies that can boost productivity without costing the environment. This is one of the challenges engineers will face in the coming century.

Green engineering also deals with industrial ecosystem, which should be similar to natural ecosystem, if not the same, so that the products become environment-friendly, which is the only way to make the product marketable and customer preferred.

In the present environment of global competition, product life time is becoming shorter. Therefore it is necessary for a company to produce a greater variety of products quickly, so as to reach the market when there is demand for the product.

A survey of over 700 companies showed that the companies which reach the market about six months late lose profits by as much as 35%. Thus, it becomes imperative for the companies to capture the market as early as possible, especially when the market is ripe for the product.

The survey also warns the companies not to reach the market before it is ready to accept the product. There are two ways to develop a product.

- The market dictates which product is going to be developed. This is called the push system of product realization because the market is pushing the company to develop the product that is required by the society.
- Even though the company enjoys good market for its products, it decides its present state is unsatisfactory and it wants to upgrade its system. This is known as pull type product realization.

It has been tested and proven time and again that the time-to-market immensely affects the success of a product. Be it craft production, mass production or the more recent production concepts like lean and agile production, the definition and verification of product and process characteristics play a very important role in product development.

IPPD is undoubtedly one of the most important developments in industry. It increases the productivity of a company as the time-to-market is shortened. Also, this is the area where the ground rules of the much-talked about lean and agile companies lie. In the study of the essential features of product and process characteristics, the following research issues are shown:

(i) The contribution of definition and verification to the product development

(ii) The four fundamental elements of **IPPD**

(iii) The techniques of **IPPD** with the help of case studies of various manufacturing firms that have benefited from **IPPD**.

Analysis of known facts and intelligent extrapolation for future with the efficient use of available resources is the essence of knowledge. While innovation coupled with knowledge wins the laurels, it is efficient management that makes all these good things happen.

The great consumer revolution of twentieth century was the result of the realization that "Value for Money" is a strategy in product development. The same philosophy of product development was not true for the then existing craft production. In the trade-off between mass production and flexible production, lean manufacturing is emerging with the inspiration of success stories from **Japan**. A virtual corporation is one that nobody dreamed of a decade ago. It is not possible to have a generic set of **IPPD** rules for all the types of organizations, but a common framework can be laid.

The relevance of re-configurability in robots and automation systems vis-à-vis agile manufacturing needs a look. Work on development of robot arms which can be configured to different types has been briefly reviewed by many authors. Extension of design theory investigated in the context of design of automation systems for agile manufacturing has also been discussed. The design of a novel dexterous manufacturing hand is already done awaiting development.

The future research in this area should concentrate on implementation issues in integrating product development

In-depth surveys of terms in various companies should be undertaken to:

- Identify and eliminate barriers for the team coordination
- Achieve communication transparency so that there is no data abstraction at any level of management

Distributed object storage and retrieval systems should be made our user-friendly so that better faster communication among teams is achieved. Software for variability reduction in processes should be developed.

In view of the rapid changes that are taking place in Indian economy due to liberalization, the manufacturing scenario is adopting itself to the changing requirements. Where the Indian industry had once enjoyed protected market, with product pricing decided on a cost plus profit basis, it has now to compete in an open market with foreign industries. One of the trend seen is that the industries are trying to provide constant innovation resulting in improved product features. This trend has resulted in industries divesting themselves of the earlier concept of vertical integration and getting their needs of component, elements, systems and subsystems from sophisticated subcontractors. The small-scale industries and ancillaries which are traditionally suppliers of simple and rough machined components, are now being encouraged to take up supply of highly engineered products. An example of this trend is the growth of auto ancillaries in India.

Small-scale industries are most suitable for agile manufacturing since they are usually without excess fat. They also have a flatter organization where personnel are empowered to take decisions without much bureaucracy. Setting up an Appropriate Technology Cell to elevate the level of operations in small-scale industries and make them suitable for incorporating the concepts of agile manufacturing is definitely needed.

19

Globalization Calls for Training of Personnel

The penetrating competition of the global industrial markets has forced the manufacturers to adopt new manufacturing techniques associating computers as the main integrating tool for improving productivity enhancement. Training is needed to improve ability, to learn new techniques and to stimulate involvement towards the goals and objectives of the firm. This envisages the need of evolving a focused approach to design training programme to evolve productivity enhancement. In the context, the modern industrial training methodologies/programmes must be studied and the feasibilities of a strategical training system approach must be found to overcome the problems faced in the training programmes.

The productivity of a country is important as it affects the inflation rates, the standard of living, employment, political power, economic power and even national security. The size and complexity of data, detailed and extensive records are needed in manufacturing systems. The principle characteristic of an information system in files of daata must be kept up-to-date. Elaborate records are needed in manufacturing systems to process, file and retrieve all the information necessary for day-to-day efficient operation of the systems to achieve higher productivity. Computer networking is a boon for the **PIS**, since data regarding productivity of any facet can be processed, stored and retrieved at greater speed, more accurately and efficiently. The computerized **PIS** is ideal, because it trades off a large initial cost and low recurring cost against a manual system. The objective of a **PIS** is to provide information for decision-making on planning, initiating, organizing and controlling the operations of the subsystems of the firm and to provide a synergistic and strategic productivity management in the process. This work is being tried out for manufacturing environment in Indian industries with optimal performance. The necessary hardware and software are selected according to the international requirements.

The other processes like shaping, planning, etc., tool materials like ceramics, diamond, etc., and different work materials can be considered and complete machining data bank can be developed. Instead of considering roughing and finishing operation, different depth of cuts and corresponding cutting parameters can be included. To reduce memory requirement, it can be interfaced with databases.

The development of an **ES** to select the optimum cutting parameters for **CNC** programme generation requires machining expertise. Thus, it avoids frequent reference to tables, textbooks and dependency on machining expertise. Production engineering is an area in which many problems would require experts to solve the typical practical problems, **ES** can be easily developed.

GT has emerged as an important technique in manufacturing engineering to cope with the fast changing industrial demands. Flexible manufacturing cells are becoming more significant to accommodate large product mix of fairly small batches of components. Formation of efficient cell is an exacting exercise which required detailed analyses of various constraints.

The development of an integral architecture to enhance a systematic methodology for **CE** has also been described by several authors. This is an attempt to assist engineering designers who need to collaborate intimately in the design of mechanical products. A significant challenge is to identify generic features and metrics that can be used to formulize mechanical engineering design processes in order that this process of product conception can be managed, improved and optimized within the constraints of a manufacturing environment.

In a CE set-up, typically all product design, process design, quality, logistics, etc., are all attempting to collaborate while using platforms and applications. Consequently, **CE** facilitates **TQM**, customer satisfaction and requirement management. However, a certain number of barriers to practice an effective **CE** in an enterprise are recognized. These barriers include complex products, processes and systems design, traditional hierarchical organizations and discipline incompatibility.

There is no unique model which permits the identification of how different engineering disciplines use analytical tools and associated results during the design of a complex system by a multidisciplinary design team. It is in partial fulfilment of the above that we propose a logically based distributed design methodology with the experience acquired from a domain specific manufacturing industry.

Product inspection or **QC** for justifying the fitness for use of the product in practice as per the needs of the end-users has been of immense importance in the present-day manufacturing-scene. Combining the state-of-the-art electronics and microcomputers with the sensor technology have added new dimension in this area. Keeping in view the present-day industrial needs, it is necessary to design and develop an on-line auto inspection system for multidimensional product feature measurement using electronic sensors and related computer based processing systems so as to evaluate the **QAP** research in the area for the suitable adaptability of such systems and concepts in an integrated manufacturing system with the provision for quickly determining the errors and hence the corrective actions.

Tool monitoring and estimation are essential for improved productivity of manufacturing systems. Multi-sensory approaches based on force vibration signals have been recognized as most potential methods for tool wear monitoring. Steady-state components of forces, dynamics of tangential forces and vibrations in the direction of main cutting force have been used for on-line tool wear monitoring and estimation in turning. Pattern recognition analysis by **LDF** approach, mathematical modelling using multiple regression analysis and a self-organizing method of modelling; **GMDH** are used to integrate information from different sensors and information of cutting conditions to obtain estimates of tool wear and tool failure. Different methods of pre-processing the forces have been attempted to determine the best methods of pre-processing the data. Various self-learning methods involved in pattern recognition, multi-regression and **GMDH** have also been analyzed.

Over the past few years, **CAE** techniques have been widely used for research in forging. The main advantage of cold forging is that the resulting product can be output to near net shape with improved mechanical properties: smooth finish, closer geometric tolerances and material saving. The traditional engineering process is complex and time-consuming. Development of an expert system for cold forging would reduce costs, assure high quality; and shorten lead times. This is just intended to present the critical issues related to the development of **KBES** and the scope of future research which will attract the attention of scientists, researchers and industrialists of the world.

20

New Aspects in the Fore

This book also illustrates several new aspects.

(i) **KBS** for optimum selection of machinability parameters like cutting tool material, tool geometry, cutting fluid, cutting speed, feed and depth of cut for manufacture of advance technology aero engine components. The system design is modular in structure. All the three modules are integrated to meet the design goals. Those modules are built with an expert system approach and each module has its own database and knowledge-base output of one module becomes input to the other module in chronological order. The cutting parameters such as speed, feed and depth of cut by satisfying the condition of minimum residual stress computation mechanism for each set of **VFD** is built in this system. This object driven expert system begins with goals and attempts to verify it. The packing fulfils the prime objective selecting the right process, right tool with right cutting parameters of the right machine with least dimensional instability on the precision machine parts.

(ii) **Selection** of suitable material handling equipment to be optimally used in typically specific conditions and material handling environment has been found to be a realistic management decision-making problem for achieving enhanced material handling effectiveness. The selection procedure of such equipment to a great extent has been found to be unstructured, characterized by an extensive domain dependent knowledge base and requires substantial human expertise to make it a candidate for the development of a KBES for fruitful selection of a user-friendly factor.

(iii) **Reduction in MLT:** Contemporary intensive market competition has motivated many manufacturing engineers and experts to develop computer controlled and automated factories with the emergence of complex and suggest remedies from the details received from shop-floor reports. Recent developments in both hardware and software of the computer industry have come a long way to make this process easier. An endeavour is made to develop an expert system using **CE** approach for a generalized manufacturing system framework named **SPM** which has been developed and successfully implemented in small size manufacturing enterprises to improve productivity. This expert system recognizes the causes of all possible impediments based on the data gathered online (common database), suggests remedial measures to reduce the **MLT** and cost incurred through **WIP** inventory. The latest developments in the field of computer technology provide an opportunity to develop expert systems which can assess the causes of impediments accurately and suggest remedial actions approximately. The work is being developed using Turbo 'C' on an IBM/PC compatible system.

(iv) **Rapid prototyping** is an emerging technology in which every individual can see the goal, the product, and can relate to the customer. Competitive manufacturing can survive by developing products from conceptual design to final product shapes in the least amount of time satisfying the design and cost-effective criterion for production. Depending upon the technology used and its complexity, projects can take from a few months to a few years to develop from the conceptual stage to the production stage. **CAD/CAM** have already provided very efficient and effective manufacturing, because of the capabilities of the technology to design and manufacture complex **3-D** shapes. It is now possible to produce any imaginable shape as required. **RPT** techniques have been developed which makes use of **CAD** data to create a photocopied image of a complex **3-D** shape.

(v) **Simulation Tool for Planning & Optimizing Manufacturing Systems and Processes:** The growing competition together with technological and organizational changes in manufacturing processes cause many companies to face the necessity for changing their research, manufacturing, and marketing strategies. Demand for newer products, shorter delivery times, quality and environment movement create severe problems for traditional manufacturing. Simulation is an effective and powerful tool which can be used advantageously in overcoming these problems of manufacturing. Attempt has been made to present the actual shape of simulation, the advantages and disadvantages of using it and the current status and future trends in this area. Efforts are also made to give readers a better understanding of stimulation to show that it is an effective, useful and very powerful tool which can be used in planning and optimizing manufacturing systems and processes.

(vi) **Future-based Modelling (Petri Net Aided Methodology):** CAPP forms the critical bridge between CAD and CAM. Generating consistent and flexible process plans to suit a dynamic production environment is an important objective. Generative CAPP systems based on Futures are being explored by researchers and practitioners. Futures represent a collection of geometrical entities in an intelligent form and hence provide information at a higher conceptual level than in traditional CAD models. As an effort in this direction, a Future-based modelling which gives necessary manufacturing related information for downstream activities of process planning has been done. A Petri Net Aided Methodology has been proposed to generate all possible alternate operation sequences.

(vii) **Automated Vision Inspection:** The role of inspection in achieving enhanced product part quality in the face of increased productivity has been found to be of utmost importance in the present-day manufacturing scenario. Automated vision inspection enables a non-contact and non-destructive system of measurement at a faster rate, thereby making 100% inspection to be accurately implemented in manufacturing systems. The modular Futures and the scope of analyses of the various product Futures for effectively describing the product shapes and recognizing the patterns by finding out suitable shape numbers are briefly dealt with. The description of objects through shape numbers has been found to be independent of the size, orientation and position of the object.

(viii) **Pattern matching through shape numbers:** The degree of similarity between the shape numbers would recognize the unknown objects with respect to the known ones. It has been observed that the modular approach of product feature description and pattern matching through shape numbers has been quite fast and flexible to be a suitable candidate for automated vision inspection. The need for broadening our vision regarding the process of technology transfer has been felt in the present environment of liberalization and increasing integration of the Indian economy with the global markets.

(ix) **Interface between Universities and SMIs:** The perspective of **SMIs** has generally been ignored in available literature as regards technology transfer. Effort has been made to assimilate this. Attempts have been made to explore the problems faced by **SMIs** in accessing their technology requirements from universities. Various existing interface models have been explored and a model appropriate to the Indian context has been proposed as an interface between universities and **SIMs**.

(x) **Information accessing from other sources:** Strategies for manufacturing need to be updated keeping in view the access to the information maintained by the involved organizations–suppliers, customers, distributors and banks. Instead of building applications to run the business of manufacturing, we need to find out whether we can access this information from other sources. If the information we need is available and cost of accessing is less than the cost of maintaining, it would be prudent to disinvest the IT and save cost with no or very little effect on our performance in manufacturing, marketing and profitability.

(xi) **COVIRDS (Conceptual Virtual Design Systems) for 3-D product Models:** Rapid generation/manipulation and visualization of three-dimensional (3-D) shapes remains a tedious task, despite the advances in the field of **CAD** and solid modelling systems. In most current CAD systems, shape designs are typically created by using combinations of keyboard and mouse inputs. Although these shape designs are typically 3-D in nature, the user inputs are restricted to 2-D devices, i.e., the interaction approaches of keyboard and mouses are restricted to 2-D. Research is going on in developing a VR-based environment, called **COVIRDS** for creating editing and visualizing three-dimensional product models.

(xii) **Reverse engineering techniques:** Reverse engineering by scanning could be used to reproduce a part without the aid of drawings of design data in a very short time. The surface data of the model to be duplicated is captured using scanning devices and the data so collected is used to build a model for generating machining programmes. In case the part has to be moulded, the model data could be used to manufacture the mould cavity directly. The captured data could also be converted into an **STL** file or **RPT**. Reverse engineering by scanning could be used as an effective tool in agile manufacturing. Some authors have dealt with the initial part building experiences with **STL**, **RPT** systems installed, mainly to cater for the needs of photoelastic stress analysis, flow visualization, investment casting and design visualization functions.

The influence of various built-in parameters like part orientation, layer thickness and support structure over the part quality and build time are explained. The process problems associated with the selection of automated build styles and support parameters are enumerated. The associated developmental efforts required for the optimal utilization of this **RPT** system are discussed with reference to the demands of photoelastic model structure.

In the current **CAD** and **CAM** systems, features are seen as the missing link between design and manufacturing. Feature Recovery, a novel concept in features research helps analyzing a part model for downstream applications like machining and featuring. By this technique, one can use attributes to store design information in a design feature model and use this information to recover machining and featuring features.

(xiii) Feature Recognition programmes: The integration of **CAD** and **CAM** largely depends on the design and development of powerful tools for design and analysis of a given part for downstream applications like machining. With current **CAD/CAM** systems using features for designing parts and assemblies, it is increasingly important to provide fast and flexible algorithms to convert design feature models into models relevant to several other domains (example, machining). Features, recognition programmes recognize the features from the pure geometric model which does not capitalize on designer's feature data. Feature recovery emphasizes the use of attributes to generate a feature-based model of a part and use design information extensively for downstream analysis. A survey of feature recognition follows.

(xiv) Feature-Based Analysis: It can be deciphered that exploiting model data at the design level, for example, hints or attributes for automatic feature-based analysis aids the recognition process. Heuristics-based **(self-learning)** volume decomposition and composition methods reduce the size of the combination problem. Machining criteria used in the form of heuristics for feature composition and evaluation helps in processing only those volumes which are feasible for machining. An attribute-based design feature model can be used to recover features for a concurrent domain of machining.

(xv) Generating Customer shrinking tool path using Voronol diagram: Most commercial **CAD/CAM** systems today are feature-based and their use involves considerable operator effort and expertise. Designers are developing an alternative approach to automatic **CAD/CAM** integration for 3-axis machining. Instead of decomposing the shape into manufacturing features, they generate tool paths directly from the shape of the workforce using accessibility as the central and over-arching constraints. Here, we concentrate on the roughing tools paths. First, the removal volume is stratified adaptively into $\frac{2l}{d}$ machining packets, each of thickness equal to or less than the maximum depth of cut. For each tool, the area of thickness equal to or less than the maximum depth of cut and the area accessible in each slice is computed incrementally. Keeping in mind the offset of the tool by its diameter, as well as the shape of the tool holder and spindle assembly, the selection of a cutting tool is optimized for the accessible area in each cut and an operation sequence is generated taking into account the time for changing a tool. Once the tools and sequence have been selected, a customer-shrinking tool path is generated using a **Voronol diagram**. The output of this system is a comprehensive set of instructions for machining the component. This information is generated directly from the CAD representation with no operator assistance.

(xvi) Study of relationship and parts as a totality in the system: Yet others have addressed complex process control applications. However, the methodology is general and can be employed in many other complex manufacturing applications. Advances in sensor technology have led to better and cheaper sensors. These advances beget more complex process control deployment of sensors. Such deployment requires new and sophisticated techniques for information analysis and fusion of information. K-system theory is a new branch of analysis that studies the relationships of parts and in totality in the system. In particular, it treats the fusion of information from multi subsystems into an overall system.

(xvii) Use of KAPES Package: CAPP package have come a long way compared to their status just a decade ago. This has been caused mainly by the phenomenal development of powerful desktops and the availability of highly reliable database management systems. KAPES, a very popular, state-of-the-art **CAPP** package, was analyzed and found to be quite effective in being able to satisfy all the standard requirements of the industry and more. The low cost of the total system makes it very attractive for even small companies which can get back their investment within 2 years. All in all, it is an ideal productivity improvement product for the manufacturing industry, especially at the present crucial juncture of worldwide economic recession.

While the concept and philosophy of 'agile manufacturing' has had its origin in **Toyota Motor of Japan**, it has since spread to all developed nations, with tremendous advantages to the consumer in the form of easy and quick availability of high quality products at very competitive prices in a variety of shapes and forms desired by the consumer. Implementation of agile manufacturing concepts have resulted in unheard of increase in productivity level as much as 300% per unit of capital employed, with cost reduction of as much to 30 to 60 per cent.

(xviii) Green Technology Environment and Ecology for Sustainability: Alongside, another revolution that has been taking place is the evolution in recent times, of what has come to be known as 'Green technologies', i.e., technologies which cause minimum or at times no pollution of the surroundings in which they operate, while at the same time releasing minimum quality of greenhouse gases such CO_2, CGC' and Methane, etc. In fact, in the coming decades, the emphasis the world over will be adopting such of those 'Green technologies' which have minimum adverse impact on environment and ecology. Such an approach has become necessary because of the tremendous havoc caused already by the industrial activity of mankind as a whole due to the release of greenhouse gases like CO_2, methane, and chlorofluorocarbons and harmful toxic substances in solid, liquid and gaseous forms to the ambient. The cumulative effect of all this has been the result of the recent phenomena of El Nino and "La Nina" which have been responsible, according to the scientists, for much of the havoc caused in the form of floods, droughts and freaky weather around the world. Adoption of "Green technologies", however, by the whole world in their value-added activities, will bring mankind closer and closer to the, concept of "Sustainable development", i.e. development without any adverse impact on environment ecology.

(xix) Computer-aided techniques for monitoring various sheet metal working: Others have presented computer-aided techniques for monitoring various sheet metal working operations. Though the technique can be applied to any type of operation, in the present work it is proposed to monitor blanking and piercing operations. The experiment consists of blanking and piercing 40 mm diameter circular holes in mild steel and brass sheets of thickness 1 mm, using dies and punchers with different clearances and edge preparations (viz., straight, single shear, double shear and negative double shear). The die is placed in a load cell (primary transducer), which is interfaced to computer through an indicator, signal conditioner, a data acquisition system and I/O card. Software compatible with the interfacing unit is loaded in computer (Pentium) for obtaining response curves of the process. Punches with different edge conditions are used to show different characteristic deviations in response curves with change in the edge condition of punches.

(xx) Control of Quality through transformation methodology: In agile manufacturing, there is rigorous control of quality; and intelligent robots having vision sensors have to be used for ensuring high quality of even minute test features of the product. Hugh transformation methodology has been suggested for developing industrial prototypes in order to achieve high-level quality control in agile manufacturing.

It is common practice to design a product before it is manufactured. During this process of design, the important characteristics of a product are specified and these are known as design specifications. The term quality or product quality refers to the degree to which the product conforms to its design specifications. When output closely conforms to the design specifications, its quality is rated as high; when output quality is traditionally rated on a continuous sequence ranging from very low to very high. In agile manufacturing, this is considered to be inadequate and there is rigorous control of the quality and measures like machine vision as CAQC are adopted in the manufacturing process. In order to be more effective, it is suggested that an intelligent robot having vision sensor may be used to ensure quality of minute features in the component being manufactured.

(xxi) Implementation of CIM: CIM in developing countries is generally implemented in two types of set-up. A new manufacturing setup is planned to operate with CIM strategy. Alternatively, an existing set up is constantly upgraded for integrating the CIM strategy to improve the productivity of existing facility. The general experience is that **TPM** is a compulsory strategy to harness the benefits of CIM in most optimal manner.

Many companies are finding that true increase in productivity and efficiency (Woman, Jones and Roos, 1990) begins with such factors as clean and efficient process, good communication infrastructure, teamwork and a constancy of sheared vision and purpose. The challenge is simply not to crank up the speed of the machines so that their outputs (per unit of time) are increased or doubled, but to change the basic machinery or process that produces those outputs. To accomplish the latter goals, today many organizations are applying **IT-based CE** techniques through benchmarking, Organizational restructuring, "Ts" renovation [e.g., Prasad, 1997] and process re-engineering [e.g. Robers, 1994].

IT-based manufacturing and CE-based synchronous organization improves efficiency through such things as systematic elimination of waste, error-proofing, just-in-time inventory, workplace organization, etc. A number of CE metrics and Agents for IT-based manufacturing or for synchronous organization have been found quite useful at customer sites **(EDS/General Motors and Delphi accounts)** for solving complex process improvements and for handling production economic problems.

The goals of manufacturing engineering are optimizing machine time, maximize production and minimize the cost. While **CNC** programming optimum machine parameters should be supplied to suit all the conditions.

(xxii) Role played by Robots in Today's Manufacturing Environments: In today's complex manufacturing environments, robotics is playing a greater role than ever. They not only assist in the production process on the assembly level, but also at more flexible levels such as part supply and distribution and beyond. No longer do we have to deal exclusively with stationary robots but mobile robotics are becoming common. The environment that such mobile robotics systems must navigate is very complex in nature and requires special arrangements to be made to allow even the most complex robotics systems to navigate or manoeuvre in them successfully and safely. The stationary, multi-axis robots commonly used for assembly tasks require little use for assembly tasks, and require little use of sensors as their environment is controlled and their task and movements are pre-programmed into the system. Experience in mobile robots in industrial surroundings must be able to react to changes in their environment, both expected and unexpected, in an intelligent way.

To accomplish this task, mobile robots must carry systems of both sensors and actuators onboard, which are designed to sample their surroundings and allow the robot to influence their surroundings.

(xxiii) Man-Machine Systems (MMS): A distinguishing feature of our modern civilization is that man rarely works without the help of machines or devices. Machines cannot continue working for a very long time without some human intervention. Today work is done by what are called **MMS**. Most industrial workers prefer performing the tasks in sitting posture and operating the machines. In every task, a combination of man and machine/tool is employed. The design of tools/machines used in the performance of the task must be suited for the users, i.e., the chair must be of correct height for the occupant, the displays must be readable, meaning thereby that the height between the user and the design of the tool would increase the efficiency in the human performance. Less attention was previously paid in this direction. But recently, the use of tools/machines is becoming complex and so the user sometimes may be positioned in hazardous environments. An attempt has been made to develop an **ES** which will guide a user who may not be very experienced. The user is required to choose from the menu as per his requirements and the system will recommend the correct route for the processing of products.

(xxiv) Design of powder metallurgy process: With the advancement of technology, a lot of process variables can be controlled precisely and as such the manufacturing process has become sophisticated and more reliable. Forging of metal powder components is one of the recent processes which depends on so many factors like powder-type selection, powder particle size selection, lubricant selection and equipment selection, etc. There are a number of independent variables that influence good quality components finally. Computer programs based on the A.I. techniques emulating the search methodology of human expert have been written to reach the final decision in the design of powder metallurgy processes.

Because of increasing complexity in process variables and its impact on productivity, the application of the knowledge-based system in determining the process selection as well as equipment selection is of vital decision to any manufacturing engineer. This expert system shall can be used as a guideline and this has been made in areas of multiple constraint satisfaction by incorporating a confidence factor. This system can be updated or extended depending on requirements from time to time.

AM is expected to proactively anticipate changes. A suitable software system like proposed **EMOS** is required to assist the agile business, which supports current and open technologies in integrating various application subsystems including planning, production, marketing, sales and services, financial management and human resources management.

The industrial world has witnessed the emergence of various new quality engineering and management approaches during different eras and in the past three decades, a number of quality gurus have emphasized the importance of infusing managerial principles in attaining higher degree of quality in their theories. The manufacturing front clearly faces another type of task. That is, there is a fast changing manufacturing approach by which varieties of products coupled with innovations are demanded within a quick span of time. At present both academic and manufacturing communities have attempted to compile this emerging concept under the technology AM (Rick 1995). It is necessary to suitably refine and modify the currently existing approaches on quality to suit the AM environment **(Mary 1997)**. Considering this as the major objective, various authors have presented exclusive outlooks literature in this direction and have evolved appropriate interpretations.

So far, the methods presented can detect distinct chains and inversions and the type of freedom. The authors believe that unless one is able to rate the chains and inversions behaviourally the purpose of identifying distinct chains, etc. is not achieved.

In order to rate the mechanism behaviourally, various methods have been tried to detect isomorphism among kinematic chains and mechanisms, many of which are based on graph theory. However, in all the reported works, mechanisms have been represented as "undirected" graphs and the flow of motion has not been considered the directed quantity. However there is a novel approach, which is both simple and reliable. The mechanism is represented by a "Direct" graph which is reliable realistic physical situations requiring digraphs. Use is made of hamming number, a concept borrowed from digital communication theory to establish numerical measures to detect hamming string, etc. These concepts have been applied in the study of 8-bar kinematic chains and mechanisms.

(xxv) QS 9000 Standard–Future Quality System Model: It is claimed that the concepts of **TPM** fundamentally aim at applying **TQM** philosophy in maintenance function. It also appraises the need for infusing quality strategies in the TPM programme. The need for incorporating a quality system and its monitoring to entrance maintenance quality is emphasized. It is pointed out that **QS 9000** standard is expected to be a future quality system model in enterprises.

(xxvi) Fuzzy Expert System for Maintenance Planners: Manufacturing concerns adopting agile manufacturing concepts consider maintenance a significant function within the overall production environment. Maintenance planning is these systems encompasses **MRO** while inventory planning, which in most cases deals with either uncertain or imprecise data. Attempt has been made by several authors to develop a Fuzzy Expert System for **MRO** inventory planning. The knowledge-base for the fuzzy expert system is acquired by the simulation of an appropriate queuing model. Multicriteria, approximate reasoning and novel research approaches have been used in this fuzzy expert system. Comparison of the fuzzy expert system and analytical models indicate a high degree of correlation. Fuzzy expert system will help maintenance planners in making better decisions.

(xxvii) Cost Estimation/Analysis by using Cost Models: Correct determination of cost components of a product is very important for any firm to stay in business. This particularly holds true in today's competition. There is no standard practice so far calculating made by a process. A cost model can be conveniently used for the purpose. It has cost analysis to be filled by the user as input to the model, and it then calculates different cost components, both in Rs/per minute and Rs/-hour. It also provides cost component graphs to facilitate quick and easy interpretation of results. Thus, it is an organized and handy tool for the management.

Implementation of **ERP** solutions is important as the process of selecting an appropriate solution. A company that intends to implement **ERP** must be crystal-clear about the full utility of the package. This determines the solution that can be selected for implementation in the organization and the methodology of implementation.

With the advent of globalization, today's manufacturers are faced with challenges to reduce manufacturing cycle time, increase productivity and reduce cost, improve product quality and provide better customer services allowing effective communication. Market demands not only force the manufacturers to be flexible to accept challenges, but also drive them to equip themselves for continuous, never-ending improvement.

The need for continuous and economical expertise which is possible only through careful exploitation of machine intelligence must be studied. The design of an intelligent system and the performance of its self-auditing module should also be considered.

(xxviii) Implementation of TQM/DSS/TQFD using Decision Support System (DSS): The principles of **TQM** envisages the development and adoption of various techniques. One among them is **QFD** which aims to translate customers' languages into technical languages. Due to **QFD's** inadequate scope, a refined technique by name **TQFD** has been applied in place. Although **TQFD** is found to be a powerful vehicle in the **TQM** journey, its implementation demands decision support. To meet this requirement, a **DSS** has also been developed. Details of the functioning of one module of this **DSS** which is named as "Action Plan" can be presented since this module assists the manufacturing managers to plan the project for effective implementation of **TQFD**.

Manufacturing today has been multi- and inter-disciplinary. It has been a business of science to integrate management, technology and workforce using information technology to produce customized products. Agility through management, technology and workforce includes object-oriented technology for quick development, ease of change and rapid deployment, message-based open applications that are reconfigurable, network-centric, reusable, interoperable and performing, etc. for implementing scaleable system software that can help the company manufacture for one with all the advantages of automation.

21
Success Stories of Agile Manufacturing

A thousand success stories surround the implementation of agile manufacturing techniques in Western companies. Those presented here exemplify corporations that are stable, sell documented, etc. like the Xerox Corporation.

21.1 SUCCESS STORIES ABROAD

21.1.1 Harley-Davidson, USA

Harley-Davidson presents the success story of an American manufacturing company able to transform itself in the nick of time. On several occasions, **Harley-Davidson** was not stable and was almost on the brink of collapse. In fact, **Harley** had such a commanding lead in the **US** heavy-motorcycle market that the company became complacent. Products were selling so well that they saw no need for innovative new designs. There was a waiting list of customers to buy **Harley-Davidson** motorcycles, and the company priced the machines according to its own needs rather than improving productivity to give customers better value. **Harley** had such brand loyalty that some customers even had their bodies tattooed with the Harley logo.

Competition in the marketplace almost annihilated the entire **Western motorcycle industry during the 1970s.** There were no motorcycle manufacturers in **Britain**, none in **France**, only one in **Germany**, and only **Harley-Davidson in the United States**. Companies like **Honda and Kawasaki** entered the market with small, inexpensive machines, established a foothold, and then began to produce large machines that directly competed with **Harley-Davidson**. **Harley** struggled for many years to stay competitive. Compared to the new Japanese machines, the Harley product was old fashioned. Yet no investment money was allotted to redesign and retool the product.

If a customer ordered a product, several weeks might lapse before that kind of product was scheduled for manufacture. When it was finally manufactured, the finished product would be delayed due to shortage of components. This situation is not atypical for small manufacturing companies making the transition from low-volume to high-volume production.

Aided by some professional consultants, the company took the classic **JIT** approach. Cause-and-effect analysis was done using fishbone charts and Pareto charts to highlight problem areas, and extensive training was put into place for all employees. Additional one-on-one education was also required for production operators who were unfamiliar with and suspicious of **JIT** manufacturing ideas. The next step was to implement a more formal inventory-control system. Previously, the company had used an informal approach to the storage of parts. Consequently, finding the right parts was often difficult, stock levels were high, and shortages were common. The new system was straightforward–single fixed locations were assigned to each item and strict controls constituted to ensure accuracy of transactions. Where possible, locking locations were situated near assembly areas. The change resulted better stock accuracy, less damage, and reduced material movement.

The introduction of mixed-model manufacturing, the first major challenge, required reducing setup times so that the product could be manufactured more frequently in smaller batches. The alteration reduced the waiting time for a period that its production run was not scheduled for some weeks and resulted in a significant increase in volume output owing to better scheduling and better real utilization of production facilities. It however resulted in increased parts shortages because the company did not have inventory control procedures in place to allow for flexibility of demand resulting in mixed-model scheduling. The solution to this problem was the use of **Kanbans**.

Kanbans were introduced gradually, starting with a few critical components and then spreading out to all manufacturing components and subassemblies. Kanban techniques were then extended to include purchased items and, with the cooperation of vendors, delivery lead times were dramatically reduced.

With these techniques in place, **Start** embarked on a formal continuous-improvement (or kaizen) approach. Employees were trained in the ideas of continuous improvement and once the programme gained momentum, an enormous number of improvements were made in quality, flexibility, cycle times, vendor relationships and product design. The benefits have been substantial. Inventory has been reduced by 60 per cent, work-in-process has been cut to maximum of 40 hours, shortages are virtually eliminated and production flexibility increased. However, the most important result has been the increase in customer satisfaction which today outstrips many competitors in the areas of flexibility and responsiveness.

21.1.2 Xerox Corporation

Xerox Corporation was an enormously successful organization throughout the 1960s and 1970s. It achieved substantial growth each year by exploiting the innovative techniques of plain paper copying so successfully that the word "Xerox" entered the English language as a verb. ("I will Xerox that for you"). In the late 1970s, however, Xerox began to see the emergence of Japanese competition as several small companies began making inroads into the copier market with small, high-quality, inexpensive copying machines. This competition was not recognized as a major threat by Xerox because the competition products were at the low end of the market, competing in the low-margin sector that Xerox felt it could afford to lose.

Unfortunately, in two or three years, Japanese companies became a recognized threat to Xerox. They began to produce mid-range copiers, Xerox's bread-and-butter business–the kind found in every office. While the highest profits were made on high-volume copiers used by professional printing organizations, its biggest market was for the midrange office copiers. Xerox had previously owned greater than 80 per cent of this market. The wolf was at the door, and Xerox had to act. The initial analysis was very discouraging to Xerox managers. The copiers being made by the Japanese companies were not only less expensive than Xerox machines, but they were also of a much higher quality. The competition delivered higher-quality designs, features, reliability, aesthetics and customer satisfaction. Competitive benchmark showed that one Japanese competitor could sell a product for less than Xerox could make an equivalent machine.

Xerox rose to the challenge. In the troubled days of the early 1980s, under the leadership of **David Kearns**, Xerox radically changed its approach to business. Management implemented JIT manufacturing techniques and introduced a companywide quality programme that impacted the entire staff from design to sales to support and administration. Xerox pioneered the use of competitive benchmarking as a tool for focusing where change is required and for monitoring the success of that change.

Xerox had notable success with vendor relationships. The vendor base was substantially reduced and many cooperative partnership arrangements were made between Xerox and its vendors. Another area of particular success was the education programme designed to bring the entire corporation into one mind on the issues of quality. From 1982 to 1987 all 100,000 Xerox employees received 48 hours of education entitled "Leadership Through Quality", which emphasized customer expectations and zero defects.

The result of these changes–very painful for those involved and especially for people whose jobs and lives were affected–is that Xerox has now gradually regained command of the market and is once again the premier plain paper copier company in the world. Xerox has also recognized that the market is changing and has embarked on a number of new ventures within its perception of document processing.

With the spread of globalization leading to increasing need for competitiveness, management of technology assumes greater emphasis. The success of an enterprise will become dependent on how much technology-driven the organization becomes and the agility of the manufacturing operations. Agile manufacturing is the science of a business system that integrates management, technology and human resource in such a manner that provides cost and time effective flexibility to the system rapidly changing market demands/expectations. Information Technology (IT) can provide solutions for all operational/disciplines.

The crucial interface for three dimensions of a manufacturing set-up, technology, management and IT, is the human resource. Because of the Indian socio-economic structure, acceptability of computer usage by a large percentage of personnel posed a challenge for the managers. Every organization, therefore, has to pay particular emphasis on **HRM** in the form of training and informal tiered interfaces to create IT acceptability. Author's study at one of the plants, with advanced technology and multiproduct/customer environment, reinforces the positive role of HRM approach for ensuring growth and for effective use of IT for enhancing productivity and agility of technology management.

21. 2 SUCCESS STORIES OF INDIAN COMPANIES

21.2.1 Maini Group India

Here, the evolution of **Maini Group** from 1973, which is a classic example of using latest concepts in manufacturing–partly by sheer coincidence and partly by meticulous design. The Group has used appropriate information technology in its journey to attain excellence in manufacture and supply of quality and reliable products. The Maini Group used to get the data processed from IBM way back in late 1970s for inventory management and annual accounts, etc. Presently the Group has over 130 PCs of different configurations with adequate modems at each location and over 50 software packages (most of them have been developed in house).

The Group has latest communication equipment and is connected through telephone, telex, fax, e-mail, and Internet. In this group, **IT** is used for MIS, Performance Budgets, Operational Plans, Variance Analysis, Costing and Estimation and CAD. Recently, the Group has installed Pro/Engineer for proactive development of new products at **Maini Materials Movement (P) Ltd.** The Group proposes to introduce **ERP** in three of the constituent companies during this year.

The demonstration deals with the skill training of penmanship. The system measures the position of the pencil of a student in 3-D space and compares those with those of the teacher and gives audible instructions.

Use of Concepts of Agile Manufacturing M/S Maini Group of Industries

Sl.No.	*Concept*	*Before*	*Measurement*	*Target*	*After*	*% Achievements*
1.	Cellular Manufacturing	0	No. of anticells	10	8	80%
2.	Lean Manufacturing	1	No. of tonnes of prodn. capacity	10T	6T	60%
3.	Continuous Improvement	5% rejection	% Rejection	0% Rejection	0.5% Rejn.	99.5%
4.	5 S	1	Housekeeping Index	Composite Index 5	4.5	90%
5.	TQM	5.8% Rejn.	% Rejection	Rejection in some cases	0% Rejn.	100% in some cases
6.	IT Integration	-	-	-	-	-
6.1	-ERP	1	No. of units	3	2	67
6.2	-CAD	3	No. of designs	18	18	100
7.	TPM	Ongoing one of the Group Company can be reports after implementation				
8.	Training Need Based	1 Manday on the average	No. of employees trained	5 Mandays/ year	4	80

21.2.2 C-MAC Centum Electronics Ltd, Bangalore

THE PROOF OF PUDDING IS IN EATING C-MAC CENTUM ELECTRONICS LTD, BANGALORE

Category	*Pre-kaizen*	*Measurement*	*post-kaizen*	*Target*	*% of Target Achived*
Production	200	Prints per hour	220	260	33%
WIP	4	Days	2	1.5	75%
Mfg. Lead Time	434	Hours	121	30	25%
Kaizens	Not Measure-ment	No/person/month	56	175	32%
Multi-Skilling on Two Critical Processes	1	No. of Technicians	4	6 Technicians to learn 2 critical operations	66 %
Housekeeping	45	%	75	90	66%
T.E.I.	Individuals Work-ing Hard		Team Working Smarter	Enjoy Work	

21.2.3 Bharath Magnetics

PROOF OF PUDDING IS IN EATING

RESULTS OF WCM / AM IMPLEMENTATION

PRODUCT : T - 30 TRANSFORMER MANUFACTURING

CATEGORY	PRE-WCM	MEASURE-MENT	POST-WCM	TARGET	% IMPROVEMENT
PRODUCTIVITY	450	No. of Pcs /Man/Month	792	650	76 %
W I P	800	No. of Pcs	100	560	88 %
INVENTORY RM & FG	2.2	Rs. Lakhs	1.2	1.65	46 %
MFG. LEAD TIME	3	DAYS	0.25	1.5	92 %
QUALITY INHOUSE	15000	ppm	2400	10500	84 %
MULTI SKILLING ONE-UP ONE DOWN	592	Points	1099	-	85 %
SPACE SAVING					
PRODUCTION AREA	560	%	360	-	36 %
STORES AREA	140	%	65	-	54 %
CUSTOMER EVALUTION	Significant improvement indicated by >100 % increase in Sales Turnover.				
T. E. I.	Employees are looking smarter and working with much less pressure / tension.				

WCM/AM = World Class Manufacturing/Agile Manufacturing

Seal: For BHARATH MAGNETICS

Sign:

21.2.4 Electro Plasts Private Limited

RESULTS OF IMPLEMENTATION OF AGILE MANUFACTURING CONCEPT ELECTRO PLASTS PRIVATE LTD.

MFRS. OF:- High Precision Injection Moulding Tools & Plastic Components, etc

Sl. No.	*Category*	*Before*	*Measurement*	*Target*	*After*	*% Achivement*
1	Plastic components manufacturing	19,000 No.	Based on maximum cycle time reqd/comp	30,000 No.	25,000 No.	83%
2	Set up time reduction	1.5 Hrs	Based on equipment utilization	0.5 Hrs	1 Hrs	50%
3	Kaizen (Continuous improvment)	5% rejection	% rejection	1 % rejection	2 % rejection	50%
4	5S implementation	1	Housekeeping index (Based on scale)	3	2	66%
5	Training Need	2 Persons	Based on Advanced technology	4	4	100%
6	CAD	9 Design	No. of design (Quarterly)	20	18	90%
				(Using Pro-E & concurrent designing)		
7	Multi-skilling on three critical basis	2	No. of Technicians	5	4	80%

Following Modules were Exhaustively Discussed in Workshops on "Agile Manufacturing to Strengthen Domestic/Global Competitiveness" and Imparted to the Participants at all the Seven Centres to Implement in Their Industries for Competitiveness

Sl. No.	CENTRES	PERIOD OF COURSE	NO.OF PART^{S}	STATISTICAL FEEDBACK			CONSOLIDATED STATISTICAL FEEDBACK OF ALL SEVEN CENTRES BY PARTICIPANTS
				EXCELL^{T}	GOOD	SATI-S^{Y}	
01	Bangalore	7th-9th June 01	20	70 %	30%	NIL	1. Systematic method in the manufacture of customized goods in industry. 2. Impressive and thought-provoking in manufacture with high technology. 3. Customer satisfaction with quality, cost-effective product as delight. 4. Latest high-technology in manufacture, i.e, process improvement, product improvement and total quality improvement. 5. Reduce unwanted cost, elimination of waste of time, material, labour, etc. 6. Total Productivity Maintenance which leads to zero defects and zero breakdown. 7. Introduction to new concept of KAN-Ban, Poka-Yoke, Kaizen, 5S Housekeeping, teamwork, multi skill, etc. which are of great importance to production, productivity, quality, etc. for competitiveness. 8. Low cost implementation of ERP and competitiveness to exploit markets. It increases the customers with more information. 9. Management, Technology and Workforce integration by IT to delight customers. 10. 18 Modules (Appendix) connected with Agile Manufacturers were dealt in detail and they are able to understand and imagine the proper course of events, during the course of manufacturer needs for the product and fulfilment of consumer requirements. 11. Discussion of manufacture of process improvement, product improvement and total quality improvement with examples. 12. The requirements namely, data input, functionality, performance, etc., contribute a cost-effective system and environmental effectiveness to fulfil the customer requirement.
02	Mysore	2nd-4th Nov.01	20	80 %	20 %	NIL	
03	Mangalore	9th–11th Nov.01	20	75%	25 %	NIL	
04	Davanagere	5th – 7th Oct 02	19	65%	35%	NIL	
05	Gulbarga	18th-20th Feb 03	25	64%	36%	NIL	
06	Bagalkot	25th-27^{th}Feb 03	27	67%	33%	NIL	
07	Shimoga	3rd-5th May 03	38	66%	34%	NIL	

References

Ansoff's, Idea of environment servicing organization, 12.1.8

Baird Michael (1978) Applications to the study of human vision

Bart Victor, Joseph B II Pine and Andrew C. Boynton, Making Mass Customization Work

Bart Victor and Andrew C. Boynton, Mass Customization Work: Creating Precision Value

Bauman, R (1990), "CALS and concurrent engineering: business strategy or tool survival?", *Aviation Week and Space Technology,* Vol. 133 pp.1-12.

Bauman, 1990, Report on CE & CALS, 4

Biren Prasad, Concurrent Engineering Wheels

B. Prasad, Hybrid re-engineering strategies for process improvement

Carey, 1992, CE coupled with automation, 4.2

C. Jackson Grayson Jr., Carla O'Dell (Hardcover, 1988), American Business

Chisholm, 1973, Challenging issues and solving them by mutual understanding, 4

Craig Giffi, Aleda V.Rotj, and Gregory M.Seal, authors of Competing in World-class Manufacturing: America's 21st Century Challenge (Business One Irwin, 1990)'

Craig Giffi, Aleda V.Rotj, and Gregory M.Seal, Competing in world-class manufacturing, 11.2.4

David Nitzan Three-Dimensional Vision Structure for Robot Applications

David Ogilvy, Creative thinking, 12.1.1

Dertouzos, Lester, and Solow, 1989, Buying power and per capita income there and in the United States

Dove, R. (1994). The Meaning of Life and the Meaning of Agile. Production Magazine, November.

Doz, 1987 Mapping the characteristics of business.

Doz, 1987, Trade-related dislocation, 9.11

Dr. Donald Pelz of Michigan University, Creative performance, 12.1.1

Druker 1975, The "business ethics" of management theory

Dyer, Charles R. (1983), Gauge Inspection Using Hough Transforms

Edward Deming, Difference between effectiveness and efficiency, 12.1.5

Einstein's, Theory of relativity and quantum mechanics revolutionized, 9.17

Ernst and young Guide, Total cost management, 10.3

E.M. Forster, Collections of essays and broadcasts Two Cheers for Democracy (1951)

Feinberg, Tanofsky and Tarrant, Managerial psychologists, Managing creative people, 12.1.1

Feinberg, Tanofsky, and Tarrant The New Psychology for Managing People

Fisher, Jain and MacDuffie, Wharton Business School report, 10.3

Gasser, L. and Majchrzak, A. (1994). ACTION Integrates Manufacturing Strategy, Design, and Planning

Goldhar, et al., 1991, Trends in business environment of companies, 4

Grayson and O'Dell, 1988, World's first productivity leaders, during 1970 or 1980, 9.9

Goldhar, J.D., Jelinek, M., Schlie, T.W. (1991), "Flexibility and competitive advantage – manufacturing becomes a service business, "*International Journal of Technology Management*, special issue on manufacturing strategy

Goldman, S.L., Nagel, R.N. and Preiss K. (1995). Agile Competitors and Virtual Organizations: Strategies for Enriching the Customer. Van Nostrand Reinhold.

Hall, 1991, Siemake & Spann, 1991, Team approach for IPPD, 4

Hayes, R.H., Wheelwright, S.C. and Clark, K.B. (1988). Dynamic Manufacturing: Creating the Learning Organization. Free Press.

Harley Davidson website: www.harley-davidson.com

Harold J. Leavitt and others Managerial Psychology: Managing Behavior in Organizations

Hll. Siemake and Spann 1991, "Phase reviews versus fast *product development*: a business case"

Hradesky's book, 1995 "The TQM Handbook," the new benchmark text for Total Quality Management implementation

Irwin, 1990, America's 21st century challenge, 11.2.4

James Shepherd Director of Research for Advanced Manufacturing Research (AMR), Boston, The cost of just entering new purchased components, 10.3

James Shepherd, The investment Strategist

Jones, 1990, Beyond the Toyota production system: the era of lean production

Jones, 1992, Lean manufacturing, 9.1

Jayesh Chakravarthi, Transitional challenges to mid-level positions, 12.1.5

Jay Forester, an MIT professor, Building national economic model based on engineering feedback theory, 9.17

Kidd, P. and Karwowski, W. *Advances in Agile Manufacturing*, Amsterdam: IOS Press, 1994, pp. 133-136.

Kidd, P.T. (1994). Agile Manufacturing: Forging New Frontiers. Addison-Wesley.

Kidd, P.T. (1995). Agile Corporations: Business Enterprises in the 21st Century-An Executive Guide. Cheshire Henbury.

Kochan, Mitchell, and Dyer, 1982, Globalization of economic activity, 9.14

Kondratieff, 1980s, Wave presage (foretell) further economic malaise (depression), 9.17

Late Edwin Tolman, Creative person loves challenges, 12.1.1

Lord Stokes 1987, the case for the defence

Mac Arthur's *Leapfrogging* (strategy)

Mac Arthur's, Strategy of leapfrogging across the Pacific, 9.11

Markov-Modulated Poisson process

Michael L. Dertouzos, Richard K. Lester, Robert M. Solow (Hardcover, 1989), Made in America
Minitzberg (1990) The manager's job: Folklore and fact
Minitzberg, 1990, Competitive environment of the future, 9.1
Mr. S.N. Mishra, President of Bharat Fritz Werner, Concurrent engineering, 12.11.4
Nam Pyo Suh, Axiomatic Design: Advances and Applications, May 2001
Nam Pyo Suh, Theory of complexity and applications
National Conference, Leadership in the Shastras, 12.1.2
Newton's, Mechanistic perspective of the world, 9.17
Nilakant, V., Ramnarayan, S. (1998), *Managing Organizational Change*
Ordway Tead, Definition of Organization, 12.1.8
Ordway Tead, Organization as a system of collaborative creation, 12.1.8
Ordway Tead, *Human Nature and Management*
Prasad, 1994, New Port News, Shipbuilding, 4
Prof. S.K. Chakraborty of IIM, Calcutta, Management by values-Towards cultural congruence, 12.1.2
Prof. S.K. Chakraborty of IIM, Calcutta "Management by Values – Towards Cultural Congruence"
Proceedings of "The National Management Convention, organized by the AIMA held, at Bangalore on April 9-10, 1993.
Rick 1995, 21st Century Manufacturing Enterprise.
Robers, 1994 Business process reengineering
Robert Hayes, a professor at the Harvard Business School, Challengers today's conventional wisdom, 9.17
Roger Nagel Lococca Institute, Four strategic principles of agility, 11.2.3
Rudyard Istvan, Sr. VP, Boston Consulting Group, 12.1.10
Sadhana Bakshi, Need for value-oriented management, 12.1.2
Savage, 1992, Computing Bayes Factors Using a Generalization of the Savage-Dickey Density Ratio
Savage, 1992, Idea transformation, 10
Shina 1991 "The relevance of concurrent engineering in industrial technology programs"
Shina, 1991, World-class stature, 4
Steven Covey, Helicopter vision, 12.1.5
Stokes, 1986, Production demand, 9.11
Tagauchi, QFD, 4.2
Thomas A. Kochan, Daniel J.B. Mitchell, Lee Dyer; Industrial relations research in the 1970s: review and appraisal
Thomas Kuhn, *The Structure of Scientific Revolutions*
Thomas Kuhn in his seminal work, The structure of scientific revolution, 9.17
Venkat Mohan of Cadis, inc., Markets parts management software, 10.3
Vernon, 1987, P. 16, Newly industrializing nationals to enter markets, 9.11

Vernon, D. 1987. "2-D Object Recognition using Partial Contours", Image and Vision Computing, Vol. 5, No. 1.

W Edwards Deming & Joseph Juran, Total Quality Management & Deming's 14 points

Womack, Jones, Roos (1990) Machine That Changed the World, (Hardcover-Oct 1990)

Womack, Jones, Roos, 1990, Lean manufacturing, 9.1

Yankelvitch Monitor survey, Customer satisfaction survey of America, 9.14

Zemake reports, 1987, U.S. Department of Transportation received number of complaints from airline passengers, 9.14

Index